THE SCIENCE OF BREATH
THE MYSTERIES OF NATURE

THE SCIENCE OF BREATH THE MYSTERIES OF NATURE

B. PATRA

GYAN

Published by

Gyan Publishing House
5, Ansari Road
Daryaganj, New Delhi-110002
Phone: 011-47034999, 9811692060
E-mail: books@gyanbooks.com

Distribution Network
gyanbooks.com
India, USA, Canada, UK, Australia, France

ISBN : 978-81-212-3795-6 (PB)
First Published, Calcutta, 1924

2nd Impression 2020

Printed at: Gyan Press, Delhi.

The Science of Breath the Mysteries of Nature
Author: B. Patra

THE OBJECT OF THIS PUBLICATION

This world is full of misfortunes and calamities, which we have to surmount. The Almighty Providence has created this material world to test His sons. Those who can stem the tide of misfortunes, calamities, and sufferings are the true sons of God. Unmixed good is rare here. We have to pick out the real good, abandoning the bad. All the imaginable good qualities are implanted within us. They require to be cultured and nourished in such a way that they will predominate over our lower animal propensities; otherwise everything will be involved in chaos and disorder.

The world is glutted with publications of various kinds. But they contain advice in too abstruse a form, producing no moral influence upon the readers, and thus men are plunged into an unfathomable abyss of sorrow, misfortune, illness, immorality, and what not.

At this critical juncture when the world is deluded with illusion, a real and lasting stimulus should be infused into the cosmopolitan to quicken his conscience to follow the right path, that he may get pure and ever-lasting happiness both here and hereafter. With this end in view, the unquestionable truths of the Aryan philosophy, the masculine morality that enables men to overcome even the most trying situations, and the way to get perfect peace and happiness are dished up to make them agreeable to the tastes of all sectarians.

Extravagance is the pervading sin of society. Now-a-days money is lavishly squandered even on the most trifling objects. Is it then too much to hope that the cosmopolitan will not at all hesitate to part with a paltry sum for the purchase of this valuable book? A perusal of it shall then assuredly convince him that the money spent by him in purchasing it was nothing compared with the rich mine of joy and knowledge lying hidden in it.

9/2, Kashi Mitter Ghat St.
Calcutta
The 3rd February, 1924

B. PATRA

INTRODUCTION

In compiling this book I am not making any claim to originality ; my main idea is to present in clear practical form some of the best known methods of breathing exercises in a cheap and handy edition for the beginners. By practice, and carefully following the immutable laws of Nature, a good grounding in the subject may be gained.

This small work, "The Mysteries of Nature," is an attempt to unravel the mysteries that shroud the whole universe. Since the creation of man, everybody has been seeking after material improvement and pleasure, totally forgetting Him who is the fountain-head of real happiness. He marries, he begets children ; in short, he gets all that he supposes would give him happiness. In this way he passes his days in search of worldly pleasures ; and when he finds that the fire of his ambition, instead of being extinguished, only blazes more and more fiercely within him ; with the lapse of time, he is mortified at heart and hankers after a companion who will be a source of everlasting joy and happiness to him.

There is, in the heart of every individual, whether rich or poor, a burning desire for communion with the Almighty, more or less developed according to his constitution or training. This desire for Holy communion is universal and is felt by every man at some or other period of his life.

Generally speaking, however, the cares, anxieties, and misfortunes of this world overwhelm the heart and eventually do away with this yearning after God. This sudden awakening, in the heart of a man, of a desire for communion with God is due to heavenly light. Those who can manage to keep this light ablaze in their hearts, find themselves slowly, and imperceptibly, attracted towards the Holy Father, the eternal reservoir of peace and happiness.

God is the spring of love and purity and always seeks a place in holy and pure hearts. In order to get Holy communion with the Universal Soul, one should first of all purify one's inner self by carefully observing the rules of the science of breath. When a man has gained a complete mastery over it, he can easily conceive the Mysterious Being within him to his unspeakable joy, when his eye of wisdom is opened by the effulgence of celestial light. Then intoxicated with the wine of eternal bliss, he begins to dance, as it were, casting aside all thoughts of material improvement. No kind of temptation can bind him again to this worldly greatness. His mind comes under the control of his will, when he can bring himself to do one thing at a time and forget everything else.

B. PATRA

PREFACE TO THE FOURTH EDITION

In preparing the fourth edition of "The Mysteries of Nature" for the press, some new subjects are added for the good of the readers. This revised edition is presented before the public, to whom, however, we must tender our heart-felt thanks for the cordial reception they so kindly extended to the third edition. Indeed, we did not at all anticipate that the previous editions would, with its manifold imperfections, run out and call for a fourth in such a brief space of time. None the less we feel it incumbent upon us to offer an apology to the readers of the last edition, whose patience was, we are afraid, taxed rather too much by reason of the numerous mistakes that, deplorably enough, bristled up and disfigured its pages, due in the main to the utmost despatch with which it had to be both compiled and published. This time, too, we have, with profound regret, to crave the indulgence of the would-be readers on account of our failure to manage for the self-same reason to give the book anything but a cursory revision, such as was possible when it was going through the press, at any rate so far as its major portion was concerned. Nevertheless we hope the edition, as it now appears, will be found to be an improvement upon the third, and will, as such, commend itself to the acceptance of the forbearing public at least as readily and as widely as the former editions did, in which case we may assure them that we shall try our level best to weed out from the book all the errors that so lamentably found a place at the outset and are still left in it, on the occasion of publishing its next edition. The extraordinary success achieved by the "Mysteries of Nature" shows how great is the demand for a cheap, concise and really practical book upon this extremely interesting subject, and we are overwhelmed with joy that we have been the source of supplying the demand of the world.

Calcutta, The 31st December 1929 The Publisher

A SHORT BIOGRAPHY

The flowers, familiar to us and which we take much care to plant around us, are the veritable sources of delight, instructions and inspiration. They delight our senses, they soothe our soul, they bloom to serve the purposes, for which they were meant by their creator and at last in withering away they do not die, but lapse into the fuller life of eternity. Such are the lives of good men. They unfold themselves into their intrinsic beauty, untrammelled by the fetters that the work-a-day world imposes upon them, to die and to revive into the glory of Heaven. The modern world, with its "sick hurry and divided aims" does not always care to take note of a life, unassuming in appearance, and poor in external achievements. But does it for that reason become the less acceptable to the Creator? Is the flower that adorns a cottage less lovely in the eyes of God, because not blooming into the garden of a millionaire it has failed to attract the eyes of the bigger man of the world ? Certainly not.

Such is the life of this friend of mine, which I am going to present to the public in the following pages. It can boast of no startles, no feature in it is bright enough to dazzle the eyes. Shorn of outward victories, it presents an appearance, simple, pure and unadorned—so uneventful in external features that one single page can exhaust it. Yet, though winning no laurel in any sphere of humdrum existence of modern life, he is out and out a warrior fighting the baser passions of his nature and conquering them on the eternal battle field of human soul. He has not fought with weapons of steel and iron but with the sword of spirit ; his enemies were not human beings of flesh and blood but the sordid part of himself that tried to clog his march into the ultimate goal of life.

Born in 1884, Baroda Prosad Patra lost his mother at the age of two and was brought up in the house of his maternal uncle at Jangalpara in the district of Hughly on the Damodar. Thus calamity was born as twin with him and his after life, as has been stated below, is one long drawn tale of unceasing adversity. Up to ten years he had never gone to school. His father, the late blind poet Durjodhan Patra, who served in the Railway department, was unable to take charge of his son, as he had to wander from place to place in obedience to the call of official authority, He read in several schools, before being finally admitted into the Morton Institution, whence he passed the Entrance Examination in the year 1901. In the meantime, another calamity was added to the list. His father lost his eye sight in 1898 and retired from service leaving his only young son at the helm of his family amidst the swelling tempest in the voyage. After his passing the Entrance Examination he was admitted into the Ripon College but adverse circumstances proved an insuperable bar to his further progress in his educational career. He was compelled to throw up his studies and after serving in some newspapers at last joined the Postal department in the year 1909. Here I was privileged to know this man and our mutual acquaintance grew into friendship.

Calamities that dissolve the mental texture of the the weaker man are apt to push the pious on the gates of Heaven. Baroda Prosad lost his worldly mother only to come closer into the embraces of the eternal Mother of the universe. When his father became blind, his caressing looks could not visit his beloved son, but Baroda Prosad felt that he was always under the eyes of his heavenly Father, who was taking care of him from above. Indeed, these adversities smoothed down the angularities of his character and formed it into a rounded equipoise. His heart became mellowed with mercy and charity. His love and sympathy flowed in incessant stream at the doors of the poor and the wretched. He knew that every morsel of food that is advanced to the lips of a hungry man is given unto God, that every kind word that is spoken to the distressed, transfuses itself into the most vital prayer at the altar of the Divinity. In accordance with this principle, he extended his charity to all, whom he thought needy and of whom he was able to do something. There appeared the divine visitation in the shape of the North Bengal flood and the Southern India flood, the cruel recollection of which still dwells in our memory, Baroda Prosad Patra could not help stretching forth his kind hand to relieve the distress of the poor and the striken. Again the people of Atghara near Baruipur post office in the district of 24 Parganas were suffering from scarcity of water and he was moved to allay their distress and he planted a tube-well there at a considerable personal sacrifice.

There is hardly a life that does not fall on evil time sometime or other and mine is no exception to that. When I was confronted with adverse circumstances, I became privileged to get a peep into the inmost core of Patra's heart. The most unstinted love and sympathy that was extended unto me, constitute a wealth to me, that is better not to be laid bare before the public. Indeed these things are too sacred for publication and I would not have been constrained to do it had not the kind hearted reading public insistently called for an account of the author's life.

In the year 1926 he visited Kashmere on the Himalayas which is known as "Heaven on Earth." Being charmed with the grandeur of the natural beauties of the Himalayas he travelled for one year throughout the unsurmountable paths of the mountain and at last reached to the Samarkhand plain. Since his return from the tour he looks very sad and pensive and always likes to be alone to enjoy the eternal bliss. In the middle of 1927 he permanently retired to pass the the rest of his life in deep meditation away from the bustles of the city. He has bequeathed all his properties by a will to be utilized in the cause of the poor after his death. He will be pleased to receive any one who wants to taste the heavenly nector in this world. His teachings are very simple and effective.

Men of this type constitute the very salt of the earth. They make the human life worth living. We cannot shut our eyes to the glaring blackness that enshrouded our life. Filfth sticks to it, sordidness eats into it. But the volume of darkness that gathers around us is dissipated by the hallow of angelic light that character of this type represents. This sustains us in our trials, braces us up with hope before the gaping teeth of adversity, their goodness strikes the balance against the evil existing in the universe. From their lives we learn that it is not all evil with us in this world and through good and evil, light and darkness, we are being conducted to the realm of permanent bliss reserved for us by our beneficent father.

SVARNAKHOLA
Faridpur
1st June, 1924

Akshoy Kumar Chakraborty

CONTENTS

Page

The Mysteries of Nature

THE SCIENCE OF BREATH

A HYGIENIC PRINCIPLE

In the days of yore the Hindu Saints (Rishis) have pondered over the long and scrutinizing investigation of the science of breath. They considered it as a separate science and dealt with it so broadly and with so much precision that it is simply wonderful. Now-a-days the younger generations, deluded by the false effulgence of foreign teachings, laugh at this exposition. The practical points of this hidden science are put before the cosmopolitan for experiment and any body can make experiments if so disposed and let the world see what immeasurable good is derived therefrom. The science of breath has been made a part of the Yoga system (concentration of mind for divine communion); in short it is a kind of stepping stone to higher communion. Any one who wishes to acquire highest pleasure and happiness in this world may be a student of this science for some few months.

A complete control over the exercises of the breath is said to secure perfect joy, health and long life. The economy of breath is regarded as the greatest of all economics. To lessen the frequency of breath, to shorten its length, to change its course in different directions, are said to be the principal means of prolonging life and preserving health and youthful vigour up to a good ripe old age. Because the breath itself is the life of man; its due regulation prolongs longevity, its waste quickens its decay. All kinds of exercises make breath flow rapidly and prolong its length. During the time of any illness or disorder of the human system, its equilibrium is disturbed. It is an axiomatic truth that the number of inhalations and exhalations of man is measured. The person whose breath (inhalation and exhalation) flows 16 to 17 times in a minute or thousand times in an hour lives six scores of years with a youthful contour and a perfectly healthy constitution. By excess and other reasons, the regularity of the breath is not kept in poise and equilibrium and the duration of life is diminished proportionately. The more the breath is regulated, the longer the life is prolonged.

The breath never flows equally through both the nostrils ; a few minute's careful examination will verify the truth of it. The breath in the human body like the tide in the sea, is affected by the movements of the sun, the moon and the other celestial bodies but specially the former, whose influence upon the tide is also the greatest. The breath comes out either through the left or the right nostril for a certain period during the day and the night and is regulated according to the wane and the increase of the moon and rarely flows through both.

The sages of ancient India are of opinion that there are three principal *Nádis* or magnetic sources in each human frame. Outside the *Merudanda* (spinal chord) there are located on the left side—the *Idá Nádi*, on the right—the *Pingalá* and within them—the *Susamná*. The *Idá* is called the moon, the *Pingalá* the sun and the Susamna the *váyu* (air).

The *Idá Nádi* flows through the left nostril, the *Pingalá* through the right and when both the nostrils are flowing, which scarcely happens, the *Susamná* is said to be flowing. There are diurnal and fortnightly variations of the breath. In each nostril the breath lasts generally for one hour. The breath changes its course 24 times in both the nostrils during one cycle of the sun. The old Rishis (sages) studied this science of breath. with minuteness. They made it a kind of barometer to gauze good or bad health before its occurrence, because improper and untimely changes of the breath are the omen of future bad health and illness. The breath was also used as a proper indicator of time. There is abundance of such persons in India who, when asked what o'clock of the day it was, could tell the correct hour of the day, by a wonderful investigation of the length of the breath and the nostril through which it was flowing at the time.

In the morning after new moon the *Idá Nádi* begins to flow *i.e.* the breath flows through the left nostril, changing its course as usual after one hour. So, on the first morning after the full moon, the Pingala-Nadi will flow. This regularity of flow of the breath in man is a sign of good luck and good health for a fortnight, while the irregularity in flow indicates the contrary. Any body can, for his own interest, change the flow of breath from one nostril to another by placing his armpit on the knee lying in a slanting posture or placing something hard under the armpit. If the left armpit is so pressed the breath will soon flow through the right nostril and vice versa.

This is also performed by stuffing the nostril with some pure old cotton.

Every thing that is good is to be done when the *Idá Nádi* flows and anything that is of a bad nature ought to be performed when the *Pingalá Nádi* flows. Some few concrete hints are given here for verification. The breath through the right nostril being *Pingalá* or the sun has the power to improve the appetite and digest the food consumed. It removes the cough, preserves the body from weakness after exercise and keeps the vitality steady. The *Idá* being the moon increases cough in the human system but is said to be a good sign for any undertaking. The bowels are naturally cleared when the *Idá Nádi* flows. Anything relating to spiritual matters is to be performed when the breath flows from both the nostrils *i.e.* when the *Susamná* flows. According to the Sastras there is not a shadow of doubt in the fact that the regulation of breath is conducive both to health and longevity.

The Yoga and Tantric Sastras attribute many things to this knowledge of the science of breath ; such as success and failure, birth and death, good and evil. In short anything that concerns human weal and woe. In the Yoga Aphorisms of Patanjala, *Pranayama* is defined as the interruption in the flow of inspiration and expiration. "Here the true *Pranayama* is accepted to be the interruption or break in the flow of the breath. The inspirations and the expirations are the functions of which the stoppage is the most effectual act for the attainment of perfect meditation. It is called *Kumbhaka* or quiescence and the longer it is sustained the more proficient becomes the Yogin. When perfect control has been attained in this respect, the Yogin can live buried under earth for months and years without performing any organic function whatever." Any one, if so disposed, can practise himself the retention of breath under the direct guidance of a good Guru (preceptor) and is sure to attain success in no time.

Among others there are two auxiliaries of the Yoga that are to be practised at the very beginning. The posture of seating (*Ashan*) becoming resolute, the other auxiliary of Yoga, the regulation of breath (*Pranayama*) and of which firmness of posture is the cause, should be practised. With the Yogin it is performed for the purpose of fixing the thinking principle to one object of concentration, as, the suppression of the functions of the thinking principle is Yoga (Divine communion). But as we are not going to deal with the philosophy of Yoga here, so no more of it now. It will be treated in other treatise.

Pranayama is to be done after being conversant with the regular posture of seating and in the way as shown in the Sastras and is to be performed regularly twice or thrice daily under the direct supervision of an able and practical preceptor. The main object of this performance is simply the studying of the mind. The minor works on Yoga and Tantras generally show the sanitary and therapeutic advantages of performing it regularly at fixed times. The Sastras are of opinion that all diseases disappear from him who devotes himself to the regulation of breath.

The usefulness of practising this science of breath from a hygienic point of view appears to have travelled to the far west and the spiritualists of America have taken it as a tried maxim of their system. Some Europeans and Americans of high intellect and culture have recoursed to practise Yoga in the same manner as shown in the Sastras. Some spiritualists practise *Pranayama* under the name of "deep breathing." Dr. Davis, a spiritualist, gives the following process for curing diseases through this agency.

"First, if your weakness be general, and the blood is loaded with cold matter, lay flat down on your back, and while breathing *deep and slow and uniformly, will yourself become healthy* in your feet and hands, in your knees and elbows, in your hips and shoulders, in your bowels and liver, in your lungs and brain, the heart will take care of itself. In cases where the weakness is generally distributed and you are required to practise (while so prostrated and respiring) is the art of concentrating your will and desires simultaneously on the extremities first; then work upward and inward progressively ; and when, in the lapse of ten minutes of steady deep breathing, you have reached the brain, repeat the process in the ascending scale, as indicated in the manner afore-mentioned."

"By this Pneumogastric treatment of yourself, you will receive spiritual strength from the air—nothing is more certain, when, by practice, you can *breathe* deeply and heroically and at the same time put your Will upon the restoration of the general system, the art of fixing your mind upon some particularly diseased part will become less and less difficult. *Consumptive persons,* by simply breathing profoundly, and willing systematically, *may enlarge their chests and lungs beyond the possibilities of disease.* Persons of cold temperature, with irregular habits and bad practices, may "right about face" and become harmonically healthy. Learn to depend upon yourself—use the infallible remedies—Nature—and, inspite of priest or doctor, you will pass from death unto life."

"Time of exercises :—In acquiring this psychological power over the destinies of your bodily state, and in becoming a self-healing Institution, whether home or abroad, it may be necessary to practise (either while on your back, or standing, or walking, or riding) perhaps three times in each twenty four hours. Never just before meals, nor soon subsequent to them; but the true time is when chylification begins; about 90 or 120 minutes after eating. The spirit-world will aid you, by forming a secret conjunction with the Pneumogastric conductor. It is certain, gentle sufferer, do not permit yourself to doubt. Nothing is too good in nature, in matter, in spirit or in truth."

PURIFICATION OF MAGNETIC SOURCES IN HUMAN SYSTEM

Those who are very curious to learn the science of breathing should first of all purify their *Nádis* i.e. the magnetic sources in human system. It is the preliminary exercise to qualify oneself for the high order of *Pranayama*, which is to be performed under the strict eye of an able and practical preceptor.

The purification of *Nádis—Idá, Pingalá* and *Susamná,* is a very easy task and it can be done in the following manner :—

You sit down erect in a steady position and keep your mouth closed. Stop the right nostril of *Pingalá Nádi* by the thumb of your right hand and gently take in breath through the left nostril and give out the breath as gently as possible immediately without keeping the breath a second in the lungs, through the right nostril by closing the left nostril.

Again inhale through the right nostril by closing the left nostril by the left-hand thumb and without keeping the breath a single second in the lungs, give it out through the left nostril by closing the right nostril with the thumb. At the beginning continue this process 3 or 4 times on each occasion and afterwards increase it to 7 or 8 times. This should be done four times in day and night ; at dawn, noon, evening and midnight. If this process is regularly carried out daily then the purification of the *Nádis* will be complete within two months and a half.

Those who are unable to perform this process four times in 24 hours should at least perform this exercise twice in the day ; once in the morning and once in the evening.

Go on practising this process for two or three months and after that you will be astonished to see its effects upon your body. Your body will become light and all sorts of idleness, lethargy and inactivity will vanish away making room for activity, youthful vigour and vivacity etc. You shall find yourself very joyous and lively. A sweet smell will come out of your body from time to time and unconscious thought-reading takes place. These subjective states take place when the *Nádi Sodhan* is fully completed and at this stage you can practise *Pranayama* under an able guidance of a preceptor.

Different postures or Asanas and exercise of muscles.

There are various ways of postures *i.e.* Asanas enumerated in the Yoga Philosophy of Patanjal. These postures are the royal road to the concentration of mind and remove germs of the diseases from the body. To keep-the body and the mind in equilibrium every one should recourse to this practice of postures or Asanas. When any one has gained a mastery on various postures he becomes proof against all the troubles of this world such as pleasures and pains, hunger and thirst, cold and heat. In fact one can keen his body and soul in peace without taking any food or water for a long time after having perfectly mastered the various postures. There are many instances of this nature but none so striking as that of Buddha in ancient times and that of De Velera, the hero of Ireland in modern. According to the western philosophers these postures are described as psychophysical exercises and are very beneficial to health. Out of the many postures only three are mentioned here : one for the exercise of all muscles, one for starting concentration exercises and the other for enjoying rest in the easiest possible way after fatigue.

Stand erect by making the feet together and **Muscle exercises** keeping the hands on the hips. Then gently and slowly lower down the body by bending the knees but keep the position of the feet intact. Resume the former position and repeat alternate raising and lowering of the body. Do not perform this exercise automatically but concentrate your mind upon the muscles in general thinking that they are all being duly put to exercise. Repeat this daily for ten minutes at the beginning and then prolong the duration as long as you are not tired.

Erect sitting or Padmasan

Sit erect with a full projection of your breast putting the right foot on the left thigh and the left foot on the right thigh by crossing the legs. Now keep the right hand on the right thigh and the left hand on the left thigh keeping the spinal chord and the head straight. Place the tongue in the roof of the mouth from where the upper set of teeth comes out. Do not keep the two rows of teeth joined together but keep them a little apart. Then gaze at the tip of the nose concentrating the mind from all worldly thoughts and cares. This well-known posture is called the Padmasan in the Yoga aphorism. This process roots out the germs of all diseases from the body if regularly practised daily for some time. At the beginning do it for ten minutes and then gradually prolong the duration. This should be performed in empty stomach or two hours after meals. The best time of performance is at dead of night when the bustles of the world subside and the nature is in her tranquility.

Resting posture or Shabasan

Simply lie down on the ground or bed on the back slackening all stiffness of every nerve and muscle like a dead body and begin to imagine within yourself that you are driving away all the weariness and troubles of your body and mind by imparting entire rest to all your limbs. By this process mental and bodily equilibrium is preserved and it is a rest both of mind and body. Practise this for 15 minutes daily after you are too much tired both mentally and physically and you will be charmed with its beneficial efficacy upon your system, when you will not be able to give up this practice whatever hindrance may stand on your daily life.

The prime factor of human life is good fresh air. Fresh air and outdoor exercises impart health, longevity and cheerfulness to a man. Sedentary habits and occupations unless counter-balanced by walking outdoors in fresh air for some hours will soon bring one to an untimely end. The psycho-physical exercises should always be performed in open air. By these exercises one's body and mind are equally developed. In it the concentration and the will-direction are simultaneously increased. Never perform these exercises mechanically but apply your mind in them. They will be very beneficial to a broken down health and mind.

Rhythmic Breathing in Pranayama

A few words are necessary on rhythmic breathing as it deals mostly with the science of breath. It is divided into three parts namely the inspiration of breath, the technical term being *Puraka*, the retention of breath is called the *Kumbhaka*, and the expiration of breath is called the *Rechaka*.

Puraka

The inspiration of breath in *Puraka* is to be done as follows :—Sit down in *Padmasan* posture. Take in the breath through the nostrils in a continuous stream slowly and without any break. It is advisable to think within during the inspiration that you are taking in the life-power or magnetism from the air and gradually fill it in the lungs.

Kumbhaka

After taking in the breath and the lungs being fully expanded for the storing up of the fresh air containing life-energy, keep it for twice as much time as you take in the inhalation of breath by closing the nostrils with your fingers.

Rechaka

It is the expiration of breath slowly and in uniformity without any break as you do in inspiration or *Puraka.* By it the lungs give out the inhaled breath thinking all the while that the impurities of inner body, sins, diseases and weakness from the system are thrown off by exhalation absorbing the life-energy from the inhaled air. The time involved in *Rechaka* is the same as in *Puraka.*

In rhythmic breathing the mind becomes tranquil and passive and it is the best time to effect auto-suggestion. Auto-suggestion generally means intimation to one's own inner self to bring about a desired changed condition, whether be it of body or of mind. Thus while you are performing Kumbhaka you should suggest to your inner self that you are drawing within you the life-energy from the atmosphere to build up a healthy constitution by surcharging the system with vital energy taken from the air. When you perform *Puraka,* mentally think that you are taking in with breath excessive quantity of Nature's ambrosia in each inhalation from the inexhaustible store of life-energy in the air. Such exercise during *Kumbhaka* while performing *Pranayama*, will develope personal magnetism, will-force and concentration.

The rhythmic breathing is concisely dealt here to give you an idea of what are called the *Puraka, Kumbhaka* and *Rechaka.* As a friend and well-wisher I advise you all that you should not jump into this practice of *Pranayama* without first learning the preliminary rules of purification of *Nádis.* As soon as the inner part of the body is purified, which one can understand from the sweet scent from the body, he may take up *Pranayama* under the direct guidance of a practical Guru i.e. preceptor.

Pranayama proper

One should not forget that in rhythmic breathing the keeping of time or rhythm in the various processes is of greatest importance. As for instance, if you require 10 seconds in the *Puraka,* you shall have to do *Kumbhaka* for 20 seconds and Rechaka for 10 seconds i. e. if it takes 10 seconds to breathe in then you shall have to retain the breath for 20 seconds in the lungs and then give out the breath in 10 seconds. The time for *Puraka* and *Rechaka* being considered as an unit may be kept constant (of course by practice it is also increased) the time for *Kumbhaka* being only increased by multiples (2, 3 or 4 times) of the *Puraka* by due prolonged practice. There are numerous sorts of *Pranayama* in the Yoga Aphorism of Patanjal. At the beginning one should not perform large number of *Pranayamas.*

Process of Pranayama proper

Sit in *Padmasan*. Close the right nostril by the thumb, take in fresh air through the left nostril completely and in the meantime think within yourself that you are drawing in the life-energy from the atmosphere for building up your body. Now stop both the nostrils with your fingers and retain the breath in the lungs for twice as long as you took in inspiration. Then close the left nostril with the thumb and exhale through the right nostril as slowly as in *Puraka*. Now breathe in through the right nostril after closing the left one by the thumb and retain for twice as above and then exhale through the left nostril. Continue the practice for several times. In order to keep the regular rhythm one should count the beads of a rosary. Cautiously fix the time of *Puraka* so that one may not gasp after performing the *Pranayama*. *Pranayama* may be performed at dawn or at an advanced period of night in a solitary and pure place in free open air.

Location of Chief Nerves

There are many people in this age who are quite ignorant of the fact that the human-breath hardly flows with equal force through both the nostrils. Generally one nostril must preponderate over the other in flow *i. e.*, one nostril has a greater and more intense flow than the other. A few days' scrutiny would entitle every one to gauze the truth of the saying and can easily say which nose flows profusely at a particular time without putting the finger near the nostrils.

There are altogether 72,000 nerves (*Nádis*) in the human system and out of these numerous nerves three are the principal nerves namely *Idá, Pingalá* and the *Susamná.* The *Idá Nádi* is called the moon and is located in the left nostril. The *Pingalá Nádi* is called the sun and its place is in the right nostril. The *Susamná Nádi* is the air (*Váyu*) and is located within the *Idá* and the *Pingalá Nádis.* When the flow of breath is entirely through one nostril or when the flow is greater through one nostril than the other then that nostril is said to be in the flow.

When the flow of breath changes from one nostril to the other, both the nostrils flow equally for some time until the flow in the other nostril is predominant. This flow of breath is known as the flow in the *Susamná Nádi.* It scarcely happens at the time of change of breath from one nostril to the other that only a few breathings take place through one nostril and few others through the other nostril, then again some more through the former and some others through the latter and so on alternately until a permanent change of flow takes place. The time varies from five to fifteen minutes.

Influence of the Nadis

The *Idá Nádi* is located towards the left side of the spinal chord, while the *Pingalá* is resting towards the right side of spinal chord. The *Idá Nádi* is favourable for good deeds and the *Pingalá* is favourable for all bad works. *Idá Nádi* is the guardian of the left side and the *Pingalá Nádi* is the guardian of the right side. *Idá Nádi* is the protector of the North and the East and the *Pingalá Nádi* that of the South and the West.

The *Susamná Nádi* is resting within the spinal chord and is like fire and death. It is the cause of all evils and failures and destroyer of everything.

Actions when the left nostril flows

When the left nostril flows one should commence works and is sure to be crowned with success. When the *Idá Nádi* flows one should begin to do good deeds of the following nature for immediate success:—Laying foundation of palaces and buildings, general study, committing to memory one's lessons of study, thinking of higher subjects, entering in any convent, starting for far-off countries, collection of materials for any good work, bestowing gifts, marriages, wearing new apparel, taking of medicines, meeting kings and high officers, making friendship, commencing new business, collecting money, entering into new houses, sowing of seeds, making treaties, cultivation of lands, giving birth of child, visiting friends, treating diseases, extending charity, gathering of corn, starting new business, setting out for holy places and pilgrimage, taking out poison from the system, riding horses, entering a new place, bringing harvest in the house. If all deeds are performed when the *Idá Nádi* flows then there will be all-round victory, success and happiness. Practice of devotion should be commenced when the two nostrils flow and the devotee is sure to get the desired results therefrom.

All kinds of unconsciousness, fever and hysteria are cured by using the medicines when the left nostril is flowing in the patient.

Let the left nose flow when one is tired, overwhelmed with grief, attacked with fever or hysteric fits, giving words of solace to others for the purpose of lessening their evil effects upon him and getting good results.

Actions when the right nostril flows

The right nostril is called *Pingalá Nádi* and it is the seat of the sun. The following nature of works should be performed when the right nostril will flow and from various practical experiences it has been ascertained that good results and success are inevitable. One can easily testify the saying by performing the following kinds of works when the right nose flows :—Making wars with enemies, hunting excursions, learning varied sciences, reading and teaching of bad sciences, drinking of wine, doing evil works, selling cattle and other commodities, climbing hills and mountains, taking medicines to cure general diseases, taking meals, writing letters, commission of theft, practising horse and cycle riding, bodily exercise, buying and selling of goods, bathing, subduing man or woman, all sorts of love-affairs, suppression of any bodily function, attraction, doing harms to others, making love with women and dining etc. If any one does the works of above nature when the right nose flows his efforts are sure to be crowned with success. He who takes his meals when the right nose flows, will never suffer from indigestion in his life. All who are suffering from dyspepsia and other bowel complaints, should rigidly follow this rule to cure the diseases radically.

When the sign of indigestion is observed due to over-meal one should lie down on the left side for nearly half an hour making the right nose flow. Let your right nose always flow when you take your meal, when you are attacked with indigestion or other ailments of the kidney and when you are engaged in coition with your wife or perform these works when the right nose flows for the purpose of attaining good results from the actions. Engaging in the lotteries, gambling, dice-playing, singing, dual fighting and conjugal felicity are sure to be successful if undertaken when the right nose is flowing.

Actions when both the nostrils flow

When the flow of breath is equal from both the nostrils the mind is in peace and perfect calmness. One desiring his own good should only engage in communion service, offer prayers and start devotional exercises for attaining success at such a time. He who performs such deeds when both the nostrils flow, will be successful in his attempts. In whatever station of life we may be placed one should always make his memory green with the fact that God is our refuge and present help in troubles. Ye inhabitants of the world ! both high and low, rich and poor should always think of Him when your both noses flow. If you can contract a habit of this you may have a direct communion with the Higher Soul and the soul within you which is really the central point of our soul, this is the fountain head of moving power and creative will.

Life is the basis of our devotion to higher spiritualism. Its aim is to attain divine life. Concentration and consecration are the beginnings, self-surrender its path ; once your feet are set upon this path, no power on earth can prevent your progress towards the goal.

Before you surrender, you must have a clear knowledge of the self. Remember—you are not merely thought, will and emotion, nor sense-enjoyment and desire, neither blood, flesh and nerves. You are in all, not different in different centres but as a single unified consciousness within and without—a totality of all actions, desires, emotions and thoughts. You contain heart, mind and intellect in yourself, you are not merely a pure naked soul, *you are in all and all are in you.*

Purification of one's soul means purification of all that it contains—the body, the mind, heart and intellect. Surrender yourself, all you possess or what you think as your belongings, in a spirit of love and devotion to Him, Who is Divine, the Infinite, the Transcendent.

Surrender your whole self without reserve, eliminate the ignorance and the limited ego-consciousness, be one with Him, transform yourself and reborn in God, the Lord of the universe. Oh the mortal children! Have this universal consciousness—the infinite power—bliss and knowledge—this is the highest fulfilment of human life—this is devotion which you should practise when the *Susamná Nádi* will flow. The doings of good or bad works should be shunned when both the nostrils are flowing equally. All kinds of actions at this time other than prayer to God, result in failure.

You should always give up the idea of starting for distant places at this moment. Inspite of this natural hinderance if anybody starts at this time success will scarcely happen but dangers and difficulties may fall on your way and even death may come from such rash actions.

During the flow of both the nostrils all individual evils may occur. Hence one should not raise any question regarding life and death, profit and loss, victory and defeat etc. at this moment.

How to change the flow from one nostril to the other

For the purpose of changing the flow from one nostril to the other, when necessary, take in fresh air through the then flowing nose by stopping the opposite one by the thumb and then take off the thumb and press the nose with the finger through which you inhaled and then exhale the air by the non-flowing side. Continue this process for some time and the change of flow will be effected on the other side of the nose.

If you lie down for a while on the side in which the nose is flowing, the change of flow occurs soon from one nose to the other. Some are of opinion that the flow of left nostril is effected by taking Ghee *i.e.* clarified butter, that of the right nose by pure *honey* whereas the *Susamná Nádi* begins to flow by taking good and sweet pomegranates.

Law of human breath

The left nostril *i.e. Idá Nádi* commences to flow at sunrise on the 1st, 2nd, 3rd, 7th, 8th, 9th, 13th, 14th and 15th of the lunar days for one hour on the lightened fortnight *i.e.*, the full moon day. Then the flow of noses changes after every alternate hour. The right nose will begin to flow at sunrise on the rest of the days *i.e.*, on the 4th, 5th, 6th, 10th, 11th and 12th of the lunar days for one hour and then the flow changes hourly from one nose to the other. Thus it is seen that in every hour the flow of breath changes from one nose to the other. But during the fortnight ending in the new moon *i.e.*, during the waning period of the moon the above rule is reversed. In other days the left nostril flows at sunrise, changing the flow hourly. When one is attacked with cold, reverse is the case.

Any disorder of flow of noses from the above general rules is a sign of disease and evil on human system.

Prediction of day's result

Carefully scrutinise at sunrise in which of the nose the breath is flowing. If the flow is found to be in accordance with the above general rules then everything is all right. If the case is quite opposite then it is a bad omen for mental agony, loss of money, frustration of desires and other kinds of losses. It has been ascertained from practical experiences that if the left nose flows fully on Monday, Wednesday, Thursday and Friday and the right nose is flowing on Tuesday, Saturday and Sunday then everything will pass in peace and weal for you but if reverse is the case then it predicts bad signs.

Fulfilment of desires

Every man in this material world has some sort of desires within him. In order to fulfil the burning desires of a man or a woman one should first touch the mouth with that hand in which side of the nose the breath is flowing after getting up from bed in the morning. This simple process will fulfil one's due desires in no time.

How to determine the good or the bad result of a fortnight before-hand

If you see that your left nostril is flowing at day-break after awaking from sleep on the first lunar day of the lightened half of the month then it will be known that there will be no disease in the system within a fortnight. If the *Suryya Nádi* (Sun) begins to flow at that time then there will be every possibility of some diseases in the body due to excessive heat in the system inspite of your caution on the matter.

Likewise you should examine your breath on the first day of the fortnight ending in the new moon after rising from sleep at sunrise. If the right nostril flows then you should be all right in a sound mind and body for the period. Inversely if the left nostril flows at the time you will be an easy prey to the diseases due to cold. To get rid of the diseases you should put some clear old cotton in the right nostril on the first lunar day of the month to get the flow from the left nostril and you should also follow the same process to get the flow of breath from the right nostril on the remaining fortnight. The above rules are as natural as water. A simple observance of the rules will clearly prove to the operator a marvellous effect.

Other important natural phenomena

If your left nostril flows on Sunday in the morning of the second lunar day of the bright half of the month you will be a gainer of many valuable things. If the same occurs on Monday then you will get perfect peace and enjoyment during the specified period. You should always bear in mind to do the acts of gift and benevolence whenever you will inhale. The Indian Yogis (devotees) have passed the opinion in one voice that the gifts which are bestowed at the time of inhaling breath have thousand times greater value and the result is a good one. In order to fulfil one's due desires he should form a habit of touching his own face by the palm of that hand in which side the nose is flowing at rising from bed early in the morning.

Laws of starting : How to start to avoid dangers, loss and other evils

In order to start for some good or bad works you should always touch the then flowing side of the body with the other hand and put forward the leg of the flowing side of the nose by striking on the ground four times if the left nostril flows and five times when the right nostril flows. To get the wonderful result of success in all matters one should honestly observe the manner of starting. Peace will supremely rule over him. No evil will fall upon him and he will come back victorious in every action.

Always start for distant places when the left nostril flows. To start for neighbouring places the flow of right nostril is good. One should take in the breath and then put the first step of the flowing side leg when ascending on some kind of vehicles or mounting a horse. One will get a wonderful success at this. One should not leave the station if he has to go towards the East or the North when the *Chandra Nádi* is flowing nor towards the South and the West when the *Suryya Nádi* is flowing. One should not at all start for any place when both the nostrils equally exhale i.e. when the *Susamná Nádi* is predominant.

Laws of success in various works

Perform all your actions proper to the *Idá Nádi* on Monday, Wednesday, Thursday and Friday when the left nostril is flowing. *Chandra Nádi* is chiefly successful in those days during the fortnight ending in the full moon of the month.

Perform all your deeds proper to the *Pingalá Nádi* on Sunday, Tuesday and Saturday when the right nostril is flowing. *Suryya Nádi* is very useful for success in those actions during the fortnight ending in the new moon.

The natural rule is that if the *Idá Nádi* flows in the morning and noon and if the *Pingalá Nádi* flows in the evening then there will be a general success and good result in every action. A contrary of this rule produces a reverse result *i.e.*. if the right nostril flows in the morning and noon and the left nostril in the evening then you will be a loser in your actions and other evils may befall you. If the right nostril flows at sunrise and if the left nostril flows at sunset consecutively for a fortnight then one is sure to be attacked with diseases. If this flow continues for two fortnights you will not gain your cause and if this occurs for three consecutive fortnights it foretells impending danger of life.

Usefulness of Nadis for special actions

The *Idá Nádi* should be used in the extraction of poisonous matters from the system. When the left nostril will flow you should begin to perform the work of removing poison. This *Nádi* should also be used in the act of devotion and prayer. In the like manner the *Pingalá Nádi* should be used in engagement *i.e.*, in love affairs, in eating, drinking, smoking and sleeping etc. The *Susamná Nádi* should be used for praying to Almighty Providence for salvation only. It is clearly shown here how the same human breath acts differently in different positions. A simple practice will produce a wonderful result and the operator will be overwhelmed with joy as he proceeds on.

Laws to give birth to sons at will

In order to beget sons at will the *Pingalá Nádi* and *Idá Nádi* should be used together. On the 5th night of the menstruation of the female when the right nostril flows in the man and the left nostril flows in the woman the two should engage in amorous action and conjoin with fervent hope of getting a son. In this respect a determined will on both the parties are urgently needed. If the right nostril of the man flows at the commencement of conjugal union and the left nostril flows just after the union, when the female will not conceive at that time.

If the conjugal union is begun when the *Susamná Nádi* flows through the right nostril for some few seconds the female will conceive a son but he will be lean, slender and defective of some organs.

Within a fortnight after the menstrual bath of a woman if a man conjoins with his wife when the right nostril is flowing profusely in him and the left nostril (profusely) flowing in her then even a sterile woman is sure to conceive. An artificial production of flow of breath in this case to suit one's purpose is always recommended.

Natural laws to remove barrenness of woman

The breath-scientists of all ages and climes are of one opinion that if the married pair lie down together or interchange love discourses or engage themselves in conjugal union when the *Pingalá Nádi* flows in man and the *Idá Nádi* flows in woman then such conjunction will cause each one to be devoted and attached to the other thereby increasing the conjugal felicity which will deepen day by day and even a stubborn woman becomes submissive to the man by this mysterious process. The old sages, the savants of the East, proved this natural phenomena long ago and commanded every married couple to observe this rule

Laws of conjugal felicity and of making a stubborn wife submissive to their great good and benefit. The Aryans always lie down keeping the wife to his left side and consequently when they will be face to face at bed, the husband has to lie down on his left side thereby causing the flow of breath in the right nostril whereas the wife has to lie down on her right side thus causing her left nostril to flow whereby the desired condition of breath is gained, which enhances mutual love and good will from day to day. Those who intend to get an everlasting and undisturbed conjugal felicity should always follow the above rules in union. In that state of breath man and woman are naturally attached to each other. The male and the female are most impressible to love in this psychological moment of the flow of breath in both the sexes.

Laws to meet an angry man

Whenever any one has to approach an angry person always start stepping the leg of the non-flowing side and on going to the angry man keep him towards the side in which there is no flow of breath in your nostril and then exchange words with him. In this way his anger will be pacified and you will gain your objects from him.

Laws to receive favours from high personages

In order to get favours from preceptors, friends, superior officers, ministers, kings, other powerful persons and the high officials always keep them towards your left side when your left nostril is flowing and ask them for favour then your efforts will be crowned with success. If an enquiry is made regarding gain, success or other good works when the left nostril is flowing, it bespeaks success and good result therefrom.

Characteristic of non-flowing nostril

The *Shunya Nádi* or the non-breathing nostril is that in which there is no flow of breath or in which the flow of breath occurs with a lesser force. If any one takes his stand in *Shunya Nádi i.e.,* keep the person of danger such as an enemy, robber, thief, armed man or angry person, towards the non-flowing side of the nose then no one will be able to do any harm to him. In winning the enemy, in dealing with the wicked, the malicious, the mischievous, the thief and an angry master do your works of vengeance or pacification after taking the above posture.

In winning your opponent in wrangles or any other person in quarrel or dispute always keep the person towards your non-flowing side of the nose in the front. Thus you will gain your object and be a victorious one in the matter. In attending for any case against an enemy in the law court start from home by stepping that leg first in which side the breath is not flowing. In the court you should stand before him in the posture as above if possible.

All kinds of work of the above nature should never be performed by keeping them in the flowing side of the nose as in that case evil will betide you and success will never be possible in your favour.

Laws of foretelling the futurity in respect of queries

Now a very important and interesting subject is to be dealt with. A perfect knowledge of which will enable everybody to have a peep into the futurity of his own and others. A constant perusal and practice with scrutiny is the stepping stone to success in this hidden art.

The calculations of futurity chiefly depend upon the four signs and data :—Position of the querist, inhalation and exhalation of breath, name or words and Tatwas. If greater number of them stands in favour the result is good and if a greater number is against favour the result is the reverse. Now the earnest and willing readers should carefully attend to the following theories to master the subject.

THEORY No 1

Readings from the position

The querist asking his questions from upwards, forwards or just from the left side of him to whom the question is put when his left nostril is flowing or the querist enquiring from downwards, backwards or from the right, when the right nostril of him to whom the question is put, is flowing—the result of such questions is good and in favour of the querist. In case of a reverse position, such as queries from the right when the left nose is flowing the result is bad and unfavourable to the querist.

THEORY No. 2

Calculation from inhalation or exhalatian of breath

If any one has commenced to ask his queries when the fortune-teller is taking the breath within, it indicates that the wishes of the querist will be successful but if the question is put when the breath is given out then it bespeaks that his purposes will not be fruitful. In questions regarding gain or loss, success or unsuccess etc, inhalation on the part of the fortune-teller will denote gain or success whereas exhalation will indicate loss or failure.

THEORY No. 3

Calculation from the words or names

In this respect there are two opinions; some count the letters of the words uttered in the questions whereas others count only the letters of the name of the person who puts the question.

If the number of the letters be found to be even on counting when the left nostril of the fortune-teller is flowing and if the number be odd when his right nostril is flowing it bespeaks success and good result. In case of reverse occurrence the fortune-teller should indicate evil and unsuccess on the part of the querist.

THEORY No. 4

Calculation from the substitute of the Tatwas

The calculation of the Tatwas is a very difficult task on the part of a beginner. Hence we mention here a suitable substitute of it.

Hold your breath after inhalation for some time and then throw a flower upwards, if it falls towards your left side when your left nose is flowing then it is all good and successful but if it falls towards that side of the nostril in which there is no flow then it denotes unsuccess and evil. This process should be performed just after the query is put. Now from the above calculations you can determine the good or evil result of almost all questions.

Those who can truly master the results and calculations of the four data will ascertain at ease the good or bad result of all queries. If greater number is in favour then the result is good and if greater number is not in favour the result will be bad. In case the numbers for and against be equal the preceeding theory should be regarded to be of more value than those following them.

THEORY No. 5

Calculation from the facing of the querist

The querist at the time of asking question faces towards the East and the North when the left nose of the fortune-teller is flowing or the querist faces towards the South and the West when the right nose is flowing would denote that the querist will be successful in his mission and purpose.

THEORY No. 6

Calculations from position

If the querist puts question standing on the left side of the fortune-teller when his left nostril is flowing it indicates good result and if he asks from your right side when your right nose is flowing the result is good as above. If the querist takes his stand (when putting question) on your non-flowing side it indicates failure and unsuccess.

If the querist makes his queries from the non-flowing side and soon after turns back to the flowing side then all his wishes are successful but if he asks the question from your flowing side and turns to the non-flowing side immediately then the result is bad.

After considering the result according to the calculations under the first four theories, see if any of the principles in theory no. 5 or 6 speaks in favour of the querist in which case they should be counted in his favour and in no case they should be counted as arguments for unfavourable judgment.

How to solve futurity and one's own query

In order to solve one's own questions you are required to put the question to anybody and then calculate the result of the query from his flow of breath. You should keep the exact number of your query. The person to whom the question is put shall be previously instructed to note the inhalation or exhalation of his breath just when you put the question. He should also note which nostril is flowing at the time. Give him a flower immediately after your query to work out theory no 4. You carefully see to which of his side the flower falls. Now getting these data from him you better proceed to note the result for him yourself and thus get your answer.

Don't lose your heart if the process seems to you a little difficult at the beginning but it will be an easy thing in course of time if the process is practised with head and heart. A constant perusal and practice of the above process will make every body a complete master of the subject.

Laws for the fortune-teller and the party

The querist should bring something—say—a flower or a piece of red paper or any other sign on perceiving which the fortune-teller should be ready to receive question and carefully note whether the first utterance of question takes place at the time of taking the breath or giving it out. The fortune-teller should also write down the exact words of the querist, if possible, and he should also work out theory No. 4 as soon as possible.

Prediction relating child-birth

In order to know beforehand whether a female or a male child will be born of any pregnant woman the following rules should be strictly observed. If the querist asks from the left side, from upwards or forwards when the *Chandra Nádi* is flowing it indicates that the woman will give birth to a female child and if the querist asks from downwards, backwards or the right side when your *Suryya Nádi* is flowing it denotes that the lady will give birth to a male child. If the query is made from the non-breathing side it indicates that the woman will not conceive. The flow of *Susamná Nádi* denotes miscarriage. There are some breath-scientists who strongly say that a query from the breathing side predicts a son and a question from the non-breathing side denotes the birth of a female child.

Queries about any disease

A question from the breathing side denotes certain cure of the disease even if the case be a complicated one. Question uttered with rough words from the breathing side indicates cure with difficulty.

A question from the breathing side by the querist while looking at any living being at the time of questioning would imply sure cure from any serious disease.

A question first uttered when the breath is being taken in it predicts cure from the disease but if the question is put when the breath is exhaled, it denotes signs of evil and hopelessness of cure.

A question from the non-breathing side denotes helplessness in the case. In order to know the strength of the evil, follow theory No. 3. A question from the non-breathing side with an odd number of letters in the question or a question put when the *Susamná Nádi* is flowing it indicates death.

If the right nostril flows continuously for a fortnight without any break from the beginning of the year, month or fortnight, it denotes death in another month.

Laws to know the sign of death

On the first day of the year, month or fortnight if the left nostril does not flow for even a few seconds but the right nostril flows continually throughout the whole day or even if just the contrary of it occurs then it predicts great evil in the near future.

He, who cannot see the front end of his own tongue, the tip of his nose, the eye-brows, also the black spot of his eyes when any corner of the eye is pressed, will certainly leave this material world within a week or so.

Laws to be a Swaryogi and how to keep everlasting youth

He, who does not worship sun in the day time nor moon at night *i. e.* who keeps the *Suryya Nádi* or the right nostril closed by old fine cotton throughout the day and the *Chandra Nádi* or the left nostril closed as above throughout the night, is a real Yogi (devotee). The simple advice herein is to keep the left nostril in flow by day and the right nostril by night. He who performs this practice is called a *Swaryogi.*

The *Swaryogi* always stops the natural flow of the one *Nádi* and changes it to the other and thereby stops decay. In this way the perpetual youth is preserved and some of them even defy death.

The above instructions should always be practised under the direct supervision of a practical Guru i. e., instructor.

Laws of deep breathing—How to keep good health

Slowly taking in of a long breath to one's utmost power and thereby filling in the lungs with pure air and then giving out that air with the same regularity of motion evacuating the lungs as completely as possible, is called the deep breathing.

There should not be any break in the stream of inhalation or exhalation. This kind of deep breathing produces a fresh and new circulation of blood throughout the system destroying the germs of the diseases in the body. In order to perform deep breathing one should always strictly follow the rules as noted below :—

Simply stand erect facing a river or any way whence pure air is flowing and also projecting the chest a little forward. Then slowly breathe in the greatest possible breath through both the nostrils fully inflating the lungs and immediately after slowly give out the breath through the nostrils exhausting the lungs as completely as possible. This process is called the simple breathing exercise.

"The waste of health is like the waste of money; it goes on almost unnoticed by the individual concerned as it is noticed as a rule too late."

Importance of health—cause of diseases—their remedies—nature cure

The above saying is only, too true. Men are apt to forget that everything else turns out to be worthless and vain in the absence of health. Food loses its taste, music loses its melody, love loses its beauty, wealth loses its pleasures and life becomes unreal and meaningless. Health is, in fact, the primary wealth of man. But humanity at large has been accustomed to think that unhealthiness is man's normal condition. Man's body is bound to be a "temple of diseases" and it is only by the patch work of drug-doctors and physicians that man can manage to make his life bearable. He may die at any moment since he is a born mortal and there is no way to prolong one's life, since diseases must cause decay. His ignorance and extreme fatalism have made him thoroughly reconciled to his present lot. This is really ridiculous. If man's rationality make him weaker than beasts and birds what is the use of such rationality.

Man has yet to fully realise that diseases are caused by transgressing the laws of nature. She has given unto man the beautiful sunshine, the sweet air, the vitalising water and her vast soil yielding vegetation, crops and fruits without any end. But unfortunately man misuses these gifts. Man has to take food to make up the loss of vitality through his day's work—but unfortunately he over-eats. He eats too much and he eats variety of dishes to pander to his taste and to stuff his stomach to its full capacity, he drinks beverages instead of pure and simple vitalising cold water, he bathes in hot water lest he should catch cold by bathing in natural cold water, he takes hard exercise to stiffen and strengthen his muscles, he kills and eats animals to feed himself, he puts on garments preventing air from blowing over his body and calls it protection from cold, he puts on shoes to coffin his feet, puts on socks to adorn his naked legs and save it from cold, puts on hats, caps and "pugris" to save his head from Nature's glorious sun-shine, rain and air. So on he goes fighting against Nature. He gets up late in the morning and spends the day in quest of earning his bread working all the time in dingy places or cooped up workshops and offices and thus goes on losing his vital force—slowly and steadily but all the time unconsciously. This is how he pays the penalty of his misused intellect and his false idea of his superiority over Nature.

We read in our old scriptures and historical books how people lived long and lived strong for ages, till the end of their lives. Unhealthiness was not known to them. Premature death of any subject was a sign of sin committed by the king. But we take these to be Grandmamma's tales. This is because in our days we do not meet with a single corroborative example. Our hospitals are full of diseased human beings—our innumerable dispensaries all over the country, our various advertisements of medicines in the newspapers, different system of drug-physicians, the Allopath, the

Kaviraj, the Hakim, etc., all tend to show that immunity from disease for human body is a myth. Our ancient books on medicines, Ayurvedic and Yunani, show that humanity was in need of drugs ever in far remote ages. That is true. But what led them to be ill, what led people to look for such drugs? This is the question which we want to deal with and we shall try to find out a sensible answer.

Stomach is the root cause of all diseases. The best cure of all sorts of bowel complaints is the proper regulation of diet. There is no such medicine in the pharmacopoeia which can radically root out the diseases of the stomach except the regulation of food. Overeating is the main cause of diseases in the system. The diet should be regulated from the very infancy. If anybody can strictly observe the following natural laws he will never witness any disease in his life :— *After forty years of age cut short your butcher's bill, give up the habit of taking sweets, give entire rest of the stomach by fasting at least twice in a month, on the fasting day do not take any food except pure water, never drink water at the time of eating but after half-an-hour of the meals, never load your stomach full with food but only three-fourth of it should be filled with food and water, never lie down after meal within half-an-hour, after night meal walk at least a mile, always take vegetable foods in abundance, drink a cup of hot water daily in the morning, take cold bath daily, be of temperate habit, practise walking in the morning daily in the fresh air, avoid taking medicines on slight illness, take medicines only when your system is actually diseased, the medicines prepared of herbs are beneficial to human system and give permanent effects.* The observance of the above laws keeps the mind and body in peace up to a good old ripe age.

Human system does not in essence differ from the system of other created beings such as the beasts and birds. The life principle works in the same way in all living beings, one rule covers all living beings without any distinction.

Any abuse of this rule makes every transgressor liable to punishment by Mother Nature in the same way as every human mother punishes the perverted and disobedient children. From Nature that punishment comes in the shape of ill-health. A growing tree begins to decay, a blooming flower suddenly fades, a sprightly kitten dozes in the morning of life, a smiling baby begins to cry, a youthful mother begins to pine away, a strong father rolls in bed with an aching head or a swollen foot. All these are due to the abuse of Nature's laws of health. The rule of life laid down by Nature has to be obeyed in order to enjoy life. But pity it is that humanity is the greatest sinner in this matter. He is the worst transgressor. We see trees to live green and fresh for centuries, we see wild animals living in full enjoyment of their life till the last day of their pretty long life and dying in the natural course of time in accordance with Nature's laws of decay. Untimely death, infant mortality, distortion of the body are seldom to be seen in the animals and birds unless they have come into contact with their master—human beings. In their wild state, Nature is their mistress and they instinctively obey her mandates and hence they remain immune from diseases to which humanity so easily is susceptible. They don't require surgeons and physicians, drugs and antiseptics as long as they have the freedom to live in accordance with their nature. Similar would have been the case with human beings if they care to use their intellect to find out the rule of life and if he could live according to Nature's demand. Artificiality has become the bane of human life and will be the chance of his redemption :—The more artificial he will be the less.

Two easy processes of ancient Indian saints' breathing exercises are given here. They are meant for improving the general health and also for destroying the germs of the chronic diseases of the system. Besides these they have manifold beneficial effects upon the human system.

1. *Shitali Process* :—Sit erect (*Padmásan* is preferable) facing towards the direction of fresh pure

air. Project your lips like that of a crow forming like a small opening just in the manner in which many persons distort their faces when whistling and then slowly inhale pure air through the mouth and then slowly breathe out through the nostrils closing the mouth. A hissing sound may be produced at the time of breathing in fresh and pure air. This simple practice not only purifies the blood of the system but radically cures spleen and chronic fever, loss of appetite, thirst and blood-poisoning and it miraculously cures colic pains. Undergo this process three or four times a day or more if you choose but never practise more than ten or twelve minutes at a time.

2. *Shitkari* Process :—Like above slowly take in pure air with a hissing sound. In this case the projection of lips is not required but the lips should be closed leaving a small opening for the passage of pure air to go inside. Give out the breaths through the nostrils as completely as possible. This may cause even the stomach and the abdomen to shrink. The practice of these twice or thrice daily will dest roy

germs of various diseases. By this process one becomes strong and beautiful driving away idleness, lethargy and drowsiness from the system.

When performing the above exercises one should pay attention to the following instructions to get the desired fruits therefrom. When taking in breath one should think within himself that he is drawing the nature's ambrosia —the universal life energy from the atmosphere and absorbing it in the system to drive away the diseases from the system. While giving out the breath imagine that each expiration is taking away a large quantity of the disease and foul matters out of the system.

Laws to remove dyspepsia

Stand erect extending the arms in a straight line with the shoulders. Then twist the body as far as practicable to the left side keeping the position of the feet stationary, next turn towards the right side as on the left side. Now make these semicircular movements of the hands and the body keeping the position of the feet as steady as possible. All the while you should concentrate upon one idea that your digestive organs are being exercised for immediate improvement.

Some minor laws

1. In case of any attack of Diarrhoea, Cholera, Dysentery and other bowel complaints which are caused by indigestion it is chiefly observed that the left nose flows during the above illness. You can cure yourself immediately by causing or changing the flow of breath through the right nostril which will act like the fire of digestion.

2. All kinds of dangers, difficulties losses etc. can fall upon you mostly when the flow of breath is through the *Susamná Nâdi* or while the flow is through the *Pingálá Nádi* but when the *Idá Nádi* flows good luck, fortune and gains of all sorts overtake you.

3. You can hardly perceive any smell when the right nostril or the *Suryya Nádi* is flowing. This *Nádi* like the sun almost separates the chemical combination which results in the effusion of smell but by the *Idá Nádi* you can completely distinguish the smell.

4. The human breath takes place nearly 21600 times throughout day and night. From the nostril it generally goes out to the length of 8 to 12 fingers. In singing and eating it goes out to the length of 16 fingers, in walking 20 fingers, in sleep 24 fingers, in sexual intercourse and in violent exercises 30 or more.

By the exercise of *Práṇáyám* the normal length of the flow of breath is shortened gradually. The devotee may lessen the length of breath to 3 or 4 fingers by the practice of rhythmic breathing and by this means one can live for a long time without breathing.

Suppose you find—say—by holding a little cotton to a distance of 12 fingers from the nostril while you are sitting in a reclining posture that your breath reaches a distance of more than 12 fingers, suppose 18 or 20 fingers, then you must be sure that your life is fast decaying. In this critical juncture a high class herbal preparation should be taken for some time in order to fill up the system with vital electricity.

5. Moreover it is often found that one particular nostril *i.e.*, the *Idá Nádi* is in flow through most of the hours by day and night the other flowing for a short time. The patient must in such a case daily close that nostril with fine old cotton for as many hours as possible. This will greatly help the curing of the disease and other treatment should also be started along with it.

6. The western scholars will be struck with wonder to learn the oriental teaching that this human machine is worked by five primary varieties of *Váyu* (air) and five secondary ones. The five primary ones are :—*Pran, Apan, Saman, Udan, Byan.* Their resting places are the heart, lower abdomen, navel, throat and the whole body respectively.

The secondary five *Váyus* (air) are *Naga, Kurma, Krikar, Devadatta* and *Dhananjaya.* Their respective functions are eructations, opening the eyes, sneezing, yawning while *Dhananjaya* pervading the whole body does not leave it even after the soul has quitted the system.

CHAPTER II

Reading of character in faces, forms and features

From the very earliest days, the science of physiognomy is believed and acted upon by all the nations of the world. It is a science based upon long experience and hard studies. It has been found that planets, stars and constellations influence very much the life and destiny of all men and women. Man is a direct image of God and all the bad and good qualities are truly implanted within us. Outer features are the true image of our inner character. In order to study this subject thoroughly one should observe the rules laid down here with scrutiny. In time he will be a best judge of human character on seeing the outward features.

THE FINGERS

There are five fingers namely (1) the thumb (2) the first finger or Polex (3) the second or middle finger or Medius (4) the third finger or Annularies and (5) the fourth or the little finger or Auricularis.

The thumb is the finger of the Venus, the first finger is that of the Jupiter, the second finger is that of the Saturn, the third finger is that of the Sun and the fourth finger is that of the Mercury.

The first joint i.e. the nail joint of the first finger is the joint of Aries, the second joint is that of Taur-usand, the third joint is that of Gemini.

The first joint of the middle finger is that of Capricorn, the second joint is that of Aquarius and the third joint is that of Pisces. The first joint of the third finger is that of Cancer, the second joint is that of Leo and the third joint is that of Sagittarius.

The roots of the fingers are known as Mounts and there are seven different Mounts of planets in the hand. The root of the little finger is the seat or mount of the Mercury. The root of the middle finger is the seat of the Saturn, the root of the first finger is that of the Jupiter and the root of the thumb is the seat or the place of the Venus.

Now let us see what we can learn from the fingers about a man's or a woman's character and destiny. We will group them as follows :—

(a) Their lengths and formation

(b) Their nails

(c) Hair on them

(d) Lines on them

THE NAILS

If the nails are broader than their lengths then that man or woman becomes argumentatives, deeply intelligent and vastly learned. He or she desires to set their authority on their relatives and neighbours and generally becomes an editor of a journal.

If a man is in the habit of biting his nails by his teeth then it is evident that his temper is very wrathful and his character is unsteady and restless.

If the nails are long and curved then he is cruel-hearted and rash in his dealings with others. If the nails are small and unclean then he is a liar and cheat and if the nails are small and black then he becomes treacherous.

If the nails are little broader, black and curved he becomes shameless and a cheat. If the nails are white and long his character is very honest and good. If the nails are very thin then he becomes weak in health and is very deceitful. If the nails are round he is addicted to pleasures.

If there is a white spot on the nail of the thumb he will be a lover and if there is a black spot he loves in vain and goes to ruin.

If there is a white spot on the nail of the first finger he becomes a rich man but if the spot is a black one then he loses money.

If there is a white spot on the nail of the middle finger he makes a voyage in the foreign countries but if the spot is a black one he is sure to die soon.

If there is a white spot on the nail of the third finger he begets another's wealth but if the spot is a black one he becomes mean-minded and earns money by dishonesty.

If there is a white spot on the nail of the little finger he becomes a lover of science and earns money by trade and commerce but if the spot is black or yellow his death is impending.

HAIR ON THE FINGERS

If there is hair on the third joints of the fingers the man's temper is wrathful and he is restless but if no hair is visible on the joints of the fingers the man becomes a coward.

LINES AND SIGNS ON FINGERS

If many lines appear on the first joints of the fingers or on any of them then they decrease the natural attributes of those joints. If these lines on the joints are curved then the result becomes quite the reverse. If a straight line joins the first and the second joints of any finger the man can wisely surpass all worldly difficulties with ease and he is never put to any embarrassment of any nature. He meets with all round happiness in all his actions.

If a small straight line is visible on each of the joints of the fingers it predicts that the man will meet a sudden death through any accident.

THE FIRST FINGER

If a straight line rising from the root of the first finger passes through the second joint and ends at the third joint then the man becomes over-talkative but wise and reasonable.

If a curved line is visible on its third joint then he becomes the owner of another's property and riches.

If a curved line is seen on its second joint he becomes a wicked man and liar ; such persons should be well guarded as they are naturally very harmful to others.

If a line encircles the first joint of the above finger just above the second joint then he becomes a weak person and he is likely to get a hurt at his head.

If two cross signs x x are visible on its second joint then he becomes the friend of a rich and well-to-do person.

If a star sign is seen on its first joint he gets much wealth and becomes a fortunate man.

If there is a star sign * on its first joint he becomes a wild-tempered and courageous man. If a line is joined with the star sign on its second joint he becomes a mild man and his temper is good and gentle.

If a star sign appears on the third joint he becomes a moral rake and is addicted to profligacy but if a half moon sign is seen on the first joint he becomes very dull and consequently undergoes many sufferings and troubles in his life.

THE SECOND FINGER

If a straight line rising from the root of this finger ends at the third joint, then he becomes victorious in all his actions and afterwards he turns to be a great and well-to-do man. If a curved line is visible on it then he surely meets death like a hero in the battle field.

If many lines, rising from its root, end at the first joint he becomes very hard-hearted, cruel and undergoes mental agony throughout his life. If few straight lines, jutting out from its third joint, end at its first joint then he gets wealth by trade and commerce in mineral substances.

If many straight lines are seen on its first joint he becomes greedy and avaricious. If many curved lines are visible on its third joint then he becomes very poor and unfortunate.

If the sign of a triangle is seen on its third joint he becomes a man of wicked and bad character and very unfortunate. If there is a cross sign on its third joint in any woman's finger then she becomes barren. If a star sign is seen on its first joint he becomes unfortunate but if a star sign is visible on the side of his finger the man gets a short life.

THE THIRD FINGER

He becomes a fortunate man in whose third finger a straight line extends from the third joint to its first joint.

If a number of lines extends from its third joint to its first joint then he undergoes a heavy loss of money through an ill-famed woman, but if there appear some straight lines on its third joint he becomes educated and happy in all his affairs.

If one of these straight lines on its third joint extends to its side it predicts that he will be a man of fame and fortune and his name will be known throughout the country. If the lines on its third joint go down to its root then be becomes very proud and earns his fortune by skilful arts.

He earns money by virtue of his good character and ability if a line goes from the third joint to its second joint. If a curved line appears on its third joint then he is saved from dangers and difficulties owing to bad influence of the planets. Misfortune befalls him if there is a half moon sign on its third joint, but if there is a cross sign on its third joint he becomes an extravagant.

THE LITTLE FINGER OR THE FOURTH FINGER

The man becomes a scientist and philanthropist if a straight line is visible extending from the third joint to the first joint ; but if three straight lines extend from its third joint to the first joint, he becomes a greedy person.

He suffers much from weak and bad health if a thick straight line is seen on its first joint. If there is a cross sign on its first joint he becomes very unfortunate and will never get any woman by marriage.

If two or three straight lines are seen on its second joint then he becomes the discoverer of many secret sciences. If the lines on its second joint are thick and irregular he becomes a man of loose character.

He becomes an orator if a straight line extends from its third joint to the second joint.

If an irregular straight line extends from its third joint to its second joint he becomes very cunning in all his affairs ; but he becomes fortunate and successful in all his undertakings if a straight line rising from its root crosses its third joint and terminates in its second joint.

The man becomes a liar and thief if a thick as well as a cross sign are visible on its third joint. He becomes an orator if a star sign is seen on its third joint; but if a straight line extends to its root it indicates that he will be a good scholar, wise and learned.

THE THUMB

If two or three straight lines extend from the first joint of the thumb to the end of its third joint he becomes a compassionate one.

He becomes the owner of wealth and fortune if a curved line is seen on its first joint; but he is loved and respected by all if there are two or three lines that rise from the root of the thumb and extend to the second joint.

If a star sign is visible on the second joint of a woman's thumb she becomes very rich. The man is sure to meet death by hanging if a line encircles the first joint of the thumb.

MEASUREMENT AND STRUCTURES OF THE FINGERS

If the first finger is like the second or the middle

The first finger finger in length then the person is proud, vain, addicted to luxury and cogitates within himself that he is very intelligent and has a sound wisdom ; but if the first finger is smaller in size than the middle finger (if it has its natural formation) then he becomes irritative in his temper and shows skilfulness in all his actions. If the first joint of this finger is longer than its other two joints he becomes very eager to talk of religious matters.

If the second joint is longer than the other two joints then the person's aims will be highly ambitious, and if the third joint is longer than the other two joints then he becomes proud and tries to exert his superiority over his subordinates.

If this finger is as long as the middle finger then he becomes a hypocrite in all his dealings. If the first joint of this finger is like a square he becomes a fine artist and paints natural sceneries of the country.

The middle finger

If the first joint of this finger is pointed it indicates that the man is hot-tempered but he becomes grave if it is like a square. If the middle finger is curved then the man becomes a traitor and murderer, but he always likes to die if the first joint is longer than its other two joints.

If its second joint is longer then the man becomes a good agriculturist and expert in machinery.

If the second joint is plump and fatty he becomes a good mathematician and loves to read religious works, but he becomes a greedy man if the third joint is long.

He becomes a wealthy person whose third finger is as long as the first. If it is as long as the middle finger he becomes a gambler, but if it is larger then he possesses natural gifts for the fine arts.

The third finger

If the third finger is like a square then he becomes a reasonable man but if it is spatulate then he becomes a good worker, orator and dancer. If its first joint is longer than the other two joints then he is very fond of fine arts; if the second joint is longer he becomes a potter and afterwards becomes very rich. If the first joint is thick and fat then he desires to learn fine arts but if the second joint is thick he becomes avaricious and is always unwilling to spend money in good causes.

The little finger

If the little finger is as long as the middle of the first joint of the second or middle finger, the man becomes fond of learning, desires to acquire knowledge and wisdom, and dives deep to find out the deep senses of the sciences and other thoughtful writings.

If this finger is unusually long then the man becomes a cheat and hypocrite but if it is very small then he becomes very wise and intelligent.

If this finger is pointed he becomes the master of secret sciences. If it is like a square then he discovers and invents many new things. If it is spatulate he becomes engaged in the study of various sciences and becomes a good orator but if its end is fatty and thick he naturally becomes a liar and thief. If the first joint is long he acquires knowledge in sciences and is a good speaker. If the second joint is long then he becomes a good business man but if the third joint is long he becomes a liar, clever and cheat.

The thumb

If the first joint of this finger is long then he becomes wise and intelligent, a man of good character, truthful and he is successful in all his undertakings. If this finger is of middle size then he secretly hatches plans to destroy his enemies but if it is small then he becomes a simple man and does not possess the power to discriminate things and puts faith in what others say.

If the first joint is small he never gets any man as a friend and naturally becomes very dishonest. If the second joint is long and thick he becomes a reasonable man but if this joint is small the result is the reverse i.e. he is wanting in common sense and intelligence.

If the first joint is long and the second joint small then he is devoid of common sense and is unable to distinguish between right and wrong. If its third joint, which is known as the place of Venus, is very high then he possesses greater portion of animal propensities—of middling high, he becomes a simple man and lover but if it is not at all high then he never becomes a lover and passes his life as a dreary one.

The natural signs are clearly written on the fingers. By a careful study of them one can read out the inner character of a man or a woman and in course of time he may be a good palmist if he can study these simple rules with whole heart and energy.

The signs or marks on the palm of a fortunate man

1. The person on whose palm there is a mark like a fish (sign of pisces) then he becomes successful in all his works; he turns to be a rich man, gets good children and passes his life in joy and happiness.

2. He is successful in any business he undertakes, on whose palm there is visible the sign of Libra and also the square sign or the sign of a thunder.

3. On whose palm there is the sign of lotus or that of Sagittarius or the sign of a big knife or a sign with eight corners he becomes a rich and happy

man. If these signs are visible on the palm of a female, she is sure to be very happy and well-to-do. If there is the sign of a lotus, she becomes a queen. If there is the sign of a bow i.e. Sagittarius he becomes a great hero ; if there is a sign of knife he becomes a powerful warrior. If there is visible a sign with a figure of eight corners he becomes the owner of many landed properties and passes his life in perfect happiness.

4. He becomes well-versed in different sciences and is regarded and respected as a wise man on whose palm the signs of wheel, conch, standard and linseed are visible.

5. He becomes a great man and acquires great fame by charitable deeds and by his devotion to religion on whose palm the sign of a trident is visible.

6. On whose palm there is the sign of an arrow he gets a kingdom, if the sign of a war-chariot with a standard is visible then the man becomes a king and good ruler.

7. If the sign of a circle or an umbrella is seen on any one's palm then he surely becomes a king and emperor. If there is any one of the above two signs, he will enjoy wealth like a king.

8. He becomes the minister of a king on whose palm the sign of mountains, human head, female organ, water pot and the wrist ornament is visible.

9. He becomes a happy man on whose palm the sign of sun, moon, the creeper, the eye, the octagon, the triangle, the temple, the horse or the elephant is visible.

10. He becomes wealthy and lives for eighty years on whose palm the sign of an umbrella, a thunderbolt etc., are visible. If these signs are seen on the palm of a woman, then she becomes the wife of a king and the mother of princes.

11. On whose palm there are the signs of a bow or that of a lotus or of a gate then the man gets many landed properties like a king and gets much wealth and fortune and lives for eighty years.

12. On whose palm there is the sign of the tail of a fish, he becomes learned and wealthy. He is certain to inherit his relative's property.

13. If there is the island sign on any one's palm, he gets learning and acquires wealth. If there is the sign of the tail of a fish he obtains sound knowledge and learning.

14. If there is the sign of one Mudra (coin) on the palm the man becomes a king. If there are two Mudras (coins) he becomes wealthy; if there are three Mudras he becomes sickly; if there are many Mudras he becomes the father of many children.

15. If there are many lines on the palm they indicate misery; if there are very few lines they denote poverty but if there are neither too many nor too few lines the man enjoys perfect peace and happiness in mind. If the lines on the palm of the hand are of red colour, the man becomes miserable and always serves another.

A FEW INDICATIONS ON THE PALM OF A WOMAN

1. The female whose palm is short, well-shaped, red in colour, not crossed by lines and contains a few broad lines distinctly marked and raised at the middle, is sure to get a good fortune ; she becomes very rich and happy.

2. If there is the sign of a fish or Pisces on a woman's palm, it predicts that she is very fortunate amongst her sex. If the sign of lotus is in her palm then she becomes a queen and her son will be a king. If on a woman's palm, there are the signs of a conch and an umbrella she becomes the mother of a king.

3. On whose palm there is the sign of an arrow she becomes a queen and on whose palm there is the sign of a cart she becomes the wife of a farmer. On whose palm a line extends from the root of the thumb to that of the little finger the woman murders her husband. A woman with such a line on the palm should never be accepted as a wife.

4. The female on whose palm there is the sign of a trident, or that of a sword becomes very famous. On whose palm there is the sign of a house, of an elephant, of an arrow, of an island, of a banner, of a garland, of a small hill, of an ear-ring, a conch, an umbrella or fish she is sure to be loved by a king. If the signs of an ear-ring and that of a wheel are visible on the palm of a female then she gets a husband from a royal family and becomes the mother of handsome children.

5. The woman, on whose palm a sign like the gate of a palace is seen, is sure to be a queen though born in the meanest family. The woman on whose palm the lines are red, distinctly marked, well-developed, bright and shining, gets much wealth and fortune. The female on whose right palm there is the sign of a scale and on whose left palm there is the sign of an elephant, a horse or a bull, becomes the wife of a merchant.

6. The woman, on whose palm the sign of the thunder-bolt or that of a palace is visible, gives birth

to a son who goes on holy pilgrimage. She becomes the wife of an agriculturist if the sign of a cart or of a deer is visible on her palm.

7. The woman, on whose palm there are signs like a jackal, a frog, a tiger, a snake, an ass, a camel and the eyes of a cat, becomes the destroyer of her husband's happiness. The woman, on whose palm the signs of ear-rings and that of a cart are visible, gets a husband who lives long and becomes the mother of many children. The woman, on whose palm the lines are found to be of red colour, becomes the wife of a king and the mother of many children.

8. The woman, on whose palm, the lines are of a copper colour and the nails are of the same hue, becomes the mother of long-lived children. The woman, on whose palm there are the signs of a lotus and that of a water-pot, becomes a queen and the mother of many sons.

9. A woman, beautiful in formation, developed in her person and walking with her full and rounded thighs and hips in the gait of a she-elephant in rut,

and possessing eyes agile and full of desire like those of a hen pigeon, intoxicated with the wine of youth, should be regarded as a fortunate lady.

A female, possessing a luxuriant growth of dark black hair, a beautiful complexion and a pair of elevated breasts pressed against each other and whose feet fall evenly on the ground and whose body is without any superfluous hair, should be deemed as a lucky and fortunate one.

10. A woman having a pendant belly, covered with rough and brown hairs should be shunned as the repository of all evils. A female who is fond of quarrelling with her neighbours and relatives or excessively greedy should be avoided as a curse.

A female whose arched eye-brows do not meet at the root of the nose is deemed as a lucky one and be accepted as a wife.

Features betray the soul of a person and a lady whose small toes do not touch the ground should be advoided as death.

SOME NATURAL SIGNS OF GOOD AND BAD FORTUNE

Tender palm like lotus petals and not perspiring, fingers adjoining each other, copper-coloured nails, beautiful ankles, well-formed head and feet, plump like the tortoise, indicates a man to be a king.

Yellow nails, rough face, high head, feet like winnowing-fan and withered toes of the feet are the marks of sorrow, poverty and bad fortune.

Thighs like the trunk of an elephant with a few hairs and one hair on the navel are the signs of a king.

He who passes urine with great sound becomes poor and those who pass urine without any sound become very rich.

Those who possess plain and even bellies become very rich and enjoy all the pleasures of the material world ; but those whose bellies are not pot-like become very poor. Those who have snake-like bellies become indigent.

The person who has got three lines on the forehead becomes very rich and gets happiness from his sons and lives for three score years. One having a line on the forehead extending to the ears lives for one hundred years in peace and happiness. One having two such lines lives for seventy years. One having two lines on the forehead, one clear and the other faint or indistinct, lives for twenty years. One who possesses short lines on the forehead lives for forty years. He who possesses broken lines on the forehead, meets with accidental death.

The maiden who possesses long and curling hair, a round face and a deep navel should be accepted as a wife. A female whose complexion is like that of gold and whose hands are red should be regarded as a true type of chastity. The maiden who has rough and uneven hair and round eyes, becomes a widow and is always unfortunate in her surroundings.

The man who possesses curling hair meets with death in the foreign countries. A knee-joint without sufficient flesh marks good luck.

Any one who passes urine with any sound, becomes poor. When urine comes out in one, two, three, four, five or six lines and falls towards the right, it indicates the sign of great wealth. Those who pass urine in a scattered way, become poor; those who pass urine in a strong flow, become happy and get good and amiable wives. When urine remains on an equal level, a man gets wife, riches and pleasures of life; but when the urine goes down, he gets only a beautiful wife, but no dowry. One who is satisfied with sexual intercourse in too short a time lives long, but he whose sexual intercourse lasts long, does not get a long life. If the nipples of the breasts are not high, the man becomes very rich and lucky; but he who has got nipples uneven, high and of yellow colour, becomes indigent.

He who possesses tender, thin and beautiful beards becomes happy and enjoys all the pleasures of life; but he who possesses thick, short and crimson-coloured beards becomes a thief. Wicked persons possess red and rugged beards.

Those who are misers have short ears. Those who have spear-like ears become wealthy and rich.

He who has got hair on the ears meets with death very soon. One who has got a long nose becomes a thief and meets death and misfortune. A straight nose with beautiful tip and small nostrils is a mark of royalty.

The good signs in woman are well formed wrists, and hands like lotuses and palms neither depressed nor very elevated. The maiden whose small toe does not touch the ground or whose thumb is bigger than the forefinger becomes unchaste. Elevated calves, sinewy, hairy or fleshy hips, pitcher-like belly, and depressed and small buttock are the marks of poverty. Short neck is the sign of poverty. If a female possesses a long forehead, she kills the younger brother of her husband ; if her belly is long, she kills her father-in-law, and if the hips are high, she kills her husband. Hairy lips are most inauspicious for the husband. Hairy breasts, rugged ears, and sharp, uneven teeth are marks of misery. If the flesh is dark-blue, it shows she will be a thief, and if it is rough, it predicts the death of her husband. A sinewy, if uneven and dry, body predicts poverty. If the upper lip is high, it shows she will be quarrelsome and harsh.

TO READ CHARACTER IN FACES AND FEATURES

HAIR

1. Hair, thin and clear, predicts a man of weak constitution, subject to various diseases.

2. Plain and smooth hair indicates a man of good understanding, condition and gentleness. Luxuriant growth of hair indicates a man of hot temper, curly and black hair indicates heat. Hair that is straight, ending like the prickles of a porcupine, indicates a dangerous and fearful man.

3. When the hair is soft and hangs down, it indicates a humid complexion. When the hair grows rapidly, it denotes that the body will soon decline to dryness and not to moisture.

4. Soft and thick hair indicates a man of much manliness and of a constitution cold and moist.

EYES

1. Little eyes, like those of a mule, indicate poor understanding and weak memory.

2. Keen and piercing eyes with drooping eyelids indicate licentiousness and wrecklessness.

3. Eyes deepening into their sockets indicate a great mind, but full of doubts.

4. Great and big eyes indicate sluggishness, boldness, lying tendency, cruelty, etc.

EYE-BROWS AND EYE-LIDS

1. When the eye-lids are long and hairy, they indicate a man of low capacity and he is cruel in all his actions with others.

2. He who possesses short and small eye-lids is regarded as wise ; but he is cruel and litigious.

3. He who possesses short and narrow eye-brows becomes wise, good-natured, but revengeful.

4. If the eye-brows are naturally folded, they indicate a shameless person, impudent and envious of another's wealth, a great but dangerous promiser.

5. He who possesses long hair on the eye-brows united and joined, is a man of simplicity.

NOSE

1. A woman who possesses a round and long nose of beautiful feature becomes wise, prudent and chaste.

2. He who has got a nose, hairy at the tip or above, is a simple-hearted man.

3. A proportionately big nose indicates a merry and peaceful man.

4. A nose that is round at the tip with small spots, indicates a proud, credulous, faithful and vain man. A woman who possesses such a nose becomes unchaste and dangerous.

5. He who has got a big and long drooping nose becomes greedy, wise, but jealous of others' actions.

6. A high nose indicates a dangerous and licentious person with lying habits.

EAR

1. Great, big and broad ears denote a simple man of weak understanding, lazy and of weak intellect.

2. Little ears signify wisdom ; but he who has got the ears well-proportioned becomes wise, discreet, honest and brave.

3. He who has long ears becomes bold, impudent, gluttonous, etc.

MOUTH

1. He who has got one lip thicker than the other is a man of weak understanding, slow to comprehend and rather foolish.

2. He who has thick, soft, small or very thin lips, becomes a fluent speaker, very talkative, full of prudence and of a good understanding, and he who has lips well-coloured and a little thick, becomes faithful and virtuous.

3. He who has thick lips becomes very simple and believes in every thing.

4. A man or a woman who possesses a small mouth becomes peaceful, faithful, fearful, eloquent and full of wisdom and learning.

5. He who has got a big and wide mouth, is a liar and a shameful, talkative, foolish, impudent and perfidious person.

BEARD

1. He who possesses no beard, but a few moustaches becomes a very bad and wicked man.

2. He who has got a fine beard and handsome and thick hair, becomes good and wise in every thing.

3. A thin, soft beard indicates a man who is very licentious and weak ; but those who have no beard are liked very much by the female sex.

CHIN

1. A woman who has a long chin becomes very irascible and unfortunate, but in the case of a man, he becomes very fickle and cannot keep anything secret.

2. A short chin indicates all kinds of malice and vice.

3. A round chin denotes a character weak and effeminate and devoid of manliness.

4. A square chin indicates manliness and signifies much courage and strength of body. A wrinkled chin indicates a weak person.

5. He who has got a crooked chin becomes treacherous and arrogant. He becomes an enemy to

all his neighbours and is puffed up with vanity. In the case of a woman, she becomes cruel, hard-hearted, mischievous, and quarrelsome.

FACE

1. A fleshy face indicates a dangerous and fearful person who is liberal, discreet, luxurious, faithful to all, unfortunate to obtain his will, but presumptuous.

2. A round face indicates a wise but cruel man.

3. A round and little face indicates a man simple, weak and of bad memory.

4. He who has got a long face is laudable in words and deeds, he is righteous, but jealous and luxurious.

5. He who has got a broad and thick face is a natural child.

6. He who has got a pale face is very sickly, and his spleen is very bad ; but he who has a white, feminine, soft, and cold face becomes tender and weak.

THE SCIENCE OF DREAM

The present generation has very little faith in dreams, although good and bad results are hidden in them. Modern science has provided something like an approximate revelation of the problem by connecting dreams in unsuspected means and influence with our waking mental life.

It is needless to discuss the physiological or psychological observations of ancient and modern times. If any body wants to discuss the subject with minuteness, let him consult the works of Maury, Wundt, Carpenter, Vollket, Tissie, Frued and Nicoll. Only the interpretation of dreams as narrated by the ancient sages in the oldest Sanskrit works is put forward here. According to the Hindu sages, dreams are neither hallucinations, illusions, impressions, ideas nor associations of the mind. Though some of the dreams are meaningless, yet there are some which indicate either good or evil to come in the future of the dreamer. It is our main object here to place before the public the true interpretations of dreams which were so long hidden in the stupendous Sanskrit works.

The dreams are divided into three classes :—
(1) good (2) evil and (3) meaningless. They are systematically described here according to the writings of the ancient obsolete books. First of all, the relation of interpretation to the time of dreaming is unfolded.

A dream which is dreamt in the first part of the night becomes realised within a year, but if in the second part of the night, it is realised in eight months, in the third quarter of the night, it is fulfilled in three months' time and dreams in the fourth part or the last quarter of the night are realised within fifteen days. Dreams dreamt in the early morning are realised that very-day. There is no meaning in day-dreams and in too short or too long dreams. If good signs are dreamt after evil signs, the result will naturally be good. The laws about dreams of the ancient sages are given below for guidance.

LAW I

If the signs of the vision agree with the dreams under A or good class, the dreamer in order to get good results, must give up sleep at any cost for the night in which he dreamt the dream.

LAW II

When the signs of the vision agree with B or the evil class, the dreamer should try his utmost to sleep in order to avoid the evil consequences of the dream.

No fixed law can be given for the meaningless dreams.

He who sleeps after dreaming good signs in his vision, does a foolish act and for this act of his foolishness he cannot derive any benefit of the good dream. Moreover, some persons meet with great evils, misfortunes and calamities if they sleep after dreaming good dreams. Quarrel, envy, animosity, loss of property, bereavement, etc., are the results of such rash action. Good dreams should not be disclosed to anybody, as in that case they turn out to be meaningless. In the case of dreams under the head B, if the Law II is not observed, the indicated evil is sure to befall the dreamer.

A
GOOD DREAMS

1. The dream of getting a flower from a pious man denotes victory, fame, happiness and wealth.

2. The dream of getting a pot or jar filled with water denotes a male child or riches.

3. The dream of a beautiful lady coming into the house with a grain-measure in her hands denotes the advent of the goddess of fortune in the family.

4. The dream of a good and beautiful female coming into the house and easing herself denotes wealth, fortune and happiness.

5. The dream of a pious man coming into the house accompanied by his wife denotes the advent of god and goddess.

6. The dream of getting rice, paddy or flower from a pious man or woman as a blessing denotes happiness, wealth and good fortune.

7. The dream of a pearl-necklace, a flower wreath or sandal paste denotes fortune and happiness.

8. The dream of getting a banner or flag, turmeric, sugar-cane or boiled rice with clarified butter denotes wealth and good fortune in all respects.

9. The dream of a pious man or woman holding an umbrella or scattering white paddy on the head denotes fortune, riches, fame and wealth like a king.

10. The dream of obtaining curd, rice or porridge on a painted and decorated chariot and wearing a white garland indicates the above results.

11. The dream of taking nectar or curd, or of getting a large ornamental vessel denotes the same result as above.

12. The dream of obtaining a book from a pious man denotes that the dreamer will be a famous poet and scholar.

13. The dream of receiving lessons and instructions like a son from his mother denotes the above result.

14. The dream of getting lessons from a pious man or obtaining a book from him denotes the same result as above.

15. The dream of getting a book indicates the result as above.

16. The dream of being initiated spiritually by a pious man or woman denotes happiness, virtue, fame, wealth and wisdom.

17. The dream of getting spiritual lessons or an idol denotes that the dreamer will become a saint.

18. The dream of obtaining blessing from a pious man or woman denotes that the dreamer will be very rich, or a famous poet.

19. The dream of getting land from a pious man adorned with white corn or paddy denotes lordship over the world.

20. The dream of travelling in heaven along with a pious man in a chariot certainly denotes long life, happiness and wealth.

21. The dream of a pious man or woman giving his or her daughter to the dreamer denotes that he will be a rich man.

22. The dream of a tank, a sea, an ocean, a river, a rivulet, a white snake or a white mountain denotes victory in all attempts of life.

23. The dream of a dead man, a happy man and a sick man denotes longevity, happiness and recovery of health from illness respectively.

24. The dream of a beautiful lady offering the dreamer her hand denotes riches and wealth like a king.

25. The dream of a girl, of a rainbow or a white cloud, or of getting a crystal necklace denotes fame to the dreamer.

26. The dream of being asked by a pious man to become his servant denotes that the dreamer will be a holy and religious man.

27. The dream of a deity, of a pious man or a woman, or of the great goddess of fortune in the guise of a milk-maid denotes wealth and good fortune.

28. The dream of a ring, a mirror, an egg, a bald head, a bow, a dress, a metal pot, a forest and a mist denotes the same result as above.

29. The above result also follows from the dream of bleeding caused by the bite of a leech, of sitting on an animal yoked to a cart, sitting within the house on a seat or bed, seeing the moon within the room or on the lap, fighting with a king, being sold as a slave, or of flying in the sky.

30. The dream of sitting with a king, of being drowned, eating the flesh of a mule, or the dream of ink, a bottle, a war, a cage, a camel and a building denotes happiness and wealth.

31. The dream of a lock, a key and of committing murder donotes fortune and wealth.

32. The dream of a dead man, or of seeing an actually dead woman alive, or of appointing a servant, also denotes the above result.

33. The dream of an unknown person cutting the hairs from the head denotes health, wealth, fortune, as also relief from distress and troubles.

34. The dream of a pen, a pen-holder, a gun, a pigeon and of falling into a well, denotes literary fame, a high post or service, good fortune, martial renown, relief from misery and destruction of enemies respectively.

35. The dream of the body attacked on all sides by worms denotes fame.

36. The dream of catching hold of a snake, taming a tiger, of fighting with a wild beast and of quarrelling denotes success in all attempts.

37. The dream of over-hearing the talk of the dead denotes realisation of hopes and aspirations, of carrying the dead body indicates immense fortune and wealth, bathing a dead body, success and victory, having meals with the dead, good fortune ; and of lying with the dead denotes longevity.

38. The dream of eating meals in a cage, of a partridge, preparing a bed and of a shining star denotes longevity on the part of the dreamer.

39. The dream of *Crocus Indicus* or *Carthumas Tictorious* or *C. Oxy-Cantha* denotes power and

fame. The vision of a she-goat indicates fame and wealth, and that of saffron, fame and learning.

40. The vision of a hat or a cap denotes power and wealth, striking an arrow at an aim indicates fame and power, and of ring-worms throughout the body, victory and fame in all attempts.

41. The vision of sacrificing a goat to a deity or of a learned man or of a sparrow indicates fame and power. The dream of an iron-house, or of a pillow denotes supernatural powers.

42. The dream of a peacock, of a duck and of bathing denotes a beautiful wife, a fickle-minded wife and a good wife respectively.

43. The dream of having a big head denotes supernatural power, while that of a small head indicates ill fame; and of a head devoured by a beast, much wealth and happiness.

44. The dream of burying a dead body denotes recovery from illness and that of fleeing from an enemy, or of a plain denotes realisation of hope and aspiration.

45. The dream of soldiers denotes heavy rain, and of sucking up the sea, good fortune and wealth.

46. The dream of two legs indicates much wealth and fortune ; of giving alms, relief from miseries and troubles ; of a tour round a village implies landed property ; of eating the head of a beast, general gain ; of the embrace of either father or mother, happiness in every respect.

47. The dream of eating roasted flesh indicates good fortune, and that of fleeing from a tiger, immense wealth.

48. The dream of a clear sky, of a burning house, of trees or plants growing on the navel, of gold ornaments, of a star, of a circle of stars, of pure water, of a friend, wheat and barley indicates fortune, weatlh or health.

49. It indicates good fortune or recovery from long illness when one dreams of crossing a river, a rivulet, a sea, an ocean, of being bitten by a many-headed reptile or serpent, leech or bee, of acquiring land, obtaining water, winning in a fight or in gam-

bling, eating raw meat or flesh or fish, bathing in blood or striking the ground with a weapon, sucking milk from the teats of a milch-cow, devouring earth, killing an enemy and of falling into fire.

50. The dream of bathing in water, falling from a cow's horn or the moon brings immense wealth and happiness like a king to the dreamer.

51. It indicates the approach of fame and wealth when one dreams of riding a bull, a horse, an elephant, of going to a palace, climbing a mountain, a hill, a tree or of having meals and shedding tears.

52. The dream of a musical instrument denotes the gain of corn-fields. The dream of being besmeared with blood or dung, of being wounded by weapons, oppressed with boils or tumours or worms denotes wealth and happiness.

52. The fulfilment of aspirations is indicated by the dream of enjoying an unmarried woman or a dead woman. The dream of enjoying a public woman signifies an early approach of marriage.

54. The dream of being moistened with urine, of taking semen or urine, of entering a city, a hell or a sea, and of drinking blood or urine is an indication of getting good news and immense wealth soon.

55. The dream of an elephant, a king, gold, a bull or a cow, a light, cooked rice, fruits, flowers, a maid, an umbrella, a banner or a flag denotes the immediate arrival of relatives as well as the acquisition of fame, good fortune and wealth.

56. The dream of a jar, or of a pot full of water, of a pious man or woman, fire, white paddy, an actor, an actress, and of a prostitute denotes fame, fortune and wealth.

57. The dream of milk or clarified butter indicates wealth and holiness.

58. The dream of taking porridge, of rice, or of drinking curd, milk, honey, clarified butter in a lotus-leaf, and of eating sweet-meats or rice-cakes denotes princely riches.

59. The dream of taking human flesh or bird's flesh denotes wealth, good news and fulfilment of aspirations.

60. The dream of swimming on a raft denotes general greatness.

61. The dream of a tree bent down with the burden of fruits is a sure indication of wealth.

62. The dream of a white snake on one's right hand denotes fame and wealth.

63. The dream of the sun, or of the moon denotes good health and freedom from all bondage.

64. The dream of a mare, a hen, or a heron is an index of sure early marriage to the dreamer ; and the dream of being bound in chains denotes fame, wealth and happiness.

65. The dream of taking rice with curd or porridge on the banks of a river or of a tank on a lotus-leaf denotes enormous wealth as befitting a king.

66. The dream of a leech, a scorpion, or of a snake denotes fame, wealth, victory, or the birth of a male-child.

67. The dream of being tortured by an animal with its horns, by a hog or a monkey denotes much wealth and landed property.

68. The dream of fish, flesh, pearl, diamond, conch-shell and sandal-wood paste yields the same result as above.

69. The dream of wine, blood, or of eating dung also produces the above result.

70. The dream of an idol, or of any god or goddess is an indication of success, fame and wealth.

71. The dream of a flowery or fruitful water-melon, or of a mango tree is an indication of gain of wealth.

72. The dream of a smoking or burning fire is an indication of wealth, intellect and good fortune.

73. The dream of a *Myrabolum or Emblemic Myrabolum*, or of a lotus is an indication of gain of riches.

74. The dream of gods, of forefathers, or aerial beings is an indication of good fortune and happiness.

75. The dream of kissing or embracing a lady dressed in white apparel and white garland and

anointed with scents, is a sure indication of good fortune and happiness in all respects.

76. The dream of enjoying a yellow woman anointed and decorated with a yellow garland denotes good fortune and happiness.

77. The dream of all kinds of white things except ashes, bones and cotton gives the same result as above.

78. The dream of a beautiful woman or of a cheerful, pious lady, adorned with gems, coming into the house is an indication of reunion with the beloved ones.

79. The dream of an angel, or of a girl of eight years old decorated with gems is an indication that the dreamer will be a famous poet and scholar.

80. The dream of getting a fruit from a cheerful, pious and contended man or woman is an indication of the birth of a male-child.

81. The dream of getting blessing from a pious man or woman denotes its realization.

82. The dream of a holy man coming into the house is an indication of good fortune, great fame, honour and happiness in every respect.

83. The dream of getting a milch-cow quite unexpectedly is a sure indication of gaining property and a beautiful obedient wife.

84. The dream of an elephant taking one up on its head by means of its trunk indicates that the dreamer will in time attain to a very high rank or position.

85. The dream of being embraced by a pious man is a sure indication of good fortune, happiness and journey to holy places.

B

EVIL DREAMS

1. The dream of black things, with the exception of cows, pious men, elephants and deities denotes misfortunes and calamities.

2. The dream of being embraced by the dead, or by an ugly man or woman is an indication of death.

3. The dream of teeth being broken or of hair falling from the head is an indication of illness or loss of wealth and property.

4. The dream of torturing an animal having horns or claws or a man or a boy is an indication of the wrath of a king or lord falling upon the dreamer.

5. The dream of a tree falling after being cut down, of a hell, of husks, or of the showering of red-hot charcoal or ashes brings misery to the dreamer.

6. The dream of a house-fall, a hill or a mountain, or a comet, or of a headless tree brings about the same result as above.

7. The dream of falling from a chariot, from a house-top, a hill, or a mountain, or a tree-top, or from the back of a bullock, a house, an elephant, or an assor a mule is an indication of an imminent danger.

8. The dream of falling into a pit full of darkness, or into a funeral pyre of charcoal or ashes is an indication of death.

9. The dream of an umbrella being forcibly snatched away from the head by a wicked man is an indication of the death of his father, or the spiritual teacher or king.

10. The dream of a terrible cow rushing out of the house with her young calf is an indication of oss of wealth and landed property.

11. The dream of being carried away, bound in hains, by the Regent of Death, or by a barbarian is an indication of death.

12. The dream of being cursed by an angry sooth-sayer, by a pious man or woman, or by a spiritual teacher is an indication of disaster and misfortune.

13. The dream of crows quarrelling and of dogs or bears falling on one's body is an indication of death.

14. The dream of being chased by an infuriated buffalo, bear, camel, hog, or an ass is an indication of approaching illness.

15. The dream of drinking lukewarm water, of a vulture and of a yellow horse is an indication

of severe illness. The vision of a shaved head brings about quarrel. The dream of being blind or deaf, or of a mouse indicates misfortunes and calamities. The vision of a dog is an indication of death, and that of an eclipse portends misery and calamity of the country wherein the dreamer lives.

16. The dream of smoke indicates danger, and that of high wind or gale, robbery. The dream of a thief is an indication of loss ; of chewing the bark of a tree or riding a goat means death ; of a mole, illness ; of digging out water, misery ; the dream of a ship, a boat or of making rope is an indication of travels in foreign countries.

17, The dream of pelting stones denotes shame and disgrace ; of illness, misfortune ; of sand, misery and loss ; of a tiger, severe illness ; of a cat, danger and robbery ; of lightning and thunder, misfortune ; of a poison, misery ; and of being born is an indication of sin and crime.

18. The dream of ascending a cloud, or of following a deceased person is an indication of death,

of going away from a strange house, illness; of writing, danger; of taking oath, misery; of the head of a dead body, or of cleansing hands, loss; of a crane, illness or death; of going to foreign countries, severe illness; of copper, loss; of kicking on the ground, misery; and the dream of taking alms indicates great misery.

19. Severe illness or death is an indication of the dream of riding a serpent or a reptile, of being embraced by an evil spirit or a ghost, of a man sitting in clay with disfigured face, or of being besmeared with mud.

20. The dream of a young elephant, a jackal, a wolf, a nude woman with dishevelled hair, a woman dressed in red garment, of a vacant house, and a broken temple indicates the same result as above.

21. The dream of reeds, trees or plants growing on the body with the exception of the navel, of eating fish, entering the body of the mother, of falling from the sun, the moon, stars or heavenly bodies, of becoming

blind, of a lamp put out, a *Bombosa Malaboriacus*, an ant-hill, an *Erythrisum Indica*, or of a flowery *Bamfivica Acaminata, Sesami seed*, eating or getting boiled rice, or the dream of drinking liquor is an indication of misfortune, calamity and illness.

22. The dream of being chased by a monster, a ghost, an evil spirit, a devil ; of being defeated in fight, of children being taken away by a nude woman or a red woman laughing uproariously is an indication of severe illness or death.

23. The dream of swooning on a bamboo reed or other bushes or plants, of falling during walk, sinking in clay, muddy water or in a dark well, drinking oil, acquiring gold, vomiting, or of being ill of diarrhoea, of quarrelling, being made a captive, losing shoes or slippers, falling of leg-skin, falling of a lamp, cracking of a hill or a mountain, entering a flower garden or a place strewn with flowers, and of entering a dense forest abounding with monkeys is a sure indication of death or severe illness or misfortune and calamity.

24. Severe illness or death comes from the dream of a man dressed in coloured attire, of a nude dark-skinned man with a rod, a singing woman, a dark-skinned woman with long hair, long nails or of a woman possessing unusually big breasts.

25. Misfortune, calamity and illness are the outcome of the vision of powdered metals, an unclean dress, plucking a lotus, killing a snake, cutting a red flower tree, killing a man, a boar, a dog, an ass, a camel, eating the flesh of unclean birds ; of angry angels, angry, pious man or woman, king or spiritual guide ; of drowning in a river or of being swept away by its current, of performing ablution in unclean, dark, or impure water, embracing a maid, committing violence upon a man ; of loss of one's own limbs ; of being ill ; of cleansing a house with brooms, of playing with monsters, devils, monkeys, bears, of man or woman dressed in red garment, of bathing in oil, riding on a cradle, getting red-hot iron, killing a horse, fall of a conch-shell, a

flag or banner, of destroying fruits or flowers, and of being swallowed up by fish.

26. The dream of a man with severed limbs, headless, disfigured or with dishevelled hair, and of a mad or a dancing man yields the results noted above.

27. The dream of fishing denotes the death of the brother of the dreamer.

28. The dream of playing on a musical instrument, of dancing or singing dressed in red garment, and of beating a drum, or of enjoyment denotes misery, misfortune and calamity.

29. The dream of a black flower or wreath of flowers, and of men armed with weapons, such as swords, javelins, spears, or arrows, is an indication of death.

30. The dream of being forsaken by an angry son or daughter of a pious man or woman is an indication of misery and misfortune.

31. The dream of a savage, or of the minister of Death with iron chains in hand is an indication of death.

32. The dream of a broken pot, a leper, a red piece of cloth, a lion. a hog, a buffalo, a complete darkness, a horrible dead body, and of the male or female organ is an indication of misfortunes, calamities and miseries.

33. The dream of a jack-fruit, a pine tree, a thorn, a crow, a bear, a monkey, an ass, and pus or filth of the human body is an indication of sudden illness.

34. The dream of beans, peas and linseed is an indication of boils or eruptions on the body.

35. The dream of the cremation ground, white things, gramineous plants and *Acasia suma* of black colour is an indication of misery, misfortune and calamity.

36. The dream of nails, charcoal and the ashes of the funeral pyre is an indication of death.

37. The dream of enjoying a female decorated with red garland is an indication of death.

38. The dream of riding on a mule, a camel, a buffalo, a deer and a chariot is an indication of death.

39. The dream of being smeared with clarified butter, milk, honey, curd and molasses is an indication of sudden illness.

40. The dream of a dead child or of a young animal, a decapitated man or beast, and of getting a necklace of bones is an indication of misfortune, calamity and misery.

41. The dream of enjoying a female of dark complexion, or decorated with a black wreath of flowers is an indication of death.

42. The dream of honey, urine, ordure, a physician and gold or silver is an indication of death within a year.

43. The dream of deities running hither and thither, singing, laughing and dancing in a place is an indication of the total annihilation of the place.

44. The dream of a female of dark complexion laughing or singing and of a dark-skinned widow is an indication of death.

45. The dream of a wild flower, a red flower, a cotton tree in flowers and of a white piece of cloth is an indication of great misery.

46. The dream of an angry, pious man or woman is an indication of death and misfortune.

47. Bereavement is implied by the dream of a naked woman, a dark-complexioned woman, a woman without nose, a low-class widow and the fruit of a palm tree.

48. Danger, misfortune, and calamity are indicated by the dream of hair, China-rose, Oleander, oil and salt.

49. Death is certain to follow from the dream of going towards the south anointed and on the back of a buffalo, a mule or a camel.

50. Loss of wealth and property and illness are the indications of the dream of tooth-ache or loose teeth.

51. The dream of laughing in joy, witnessing a marriage ceremony, or dance, or listening to sweet music is an indication of sure misfortune to the dreamer.

C

GROUNDLESS DREAMS

Visions of idiots, of the decrepit, of a person oppressed with the suppression of the calls of nature and persons with dishevelled hair are meaningless.

Visions of anxious, ill, tortured and frightened persons are also meaningless.

Visions of things previously seen, or heard of, or enquired about, premeditated, or previously imagined are meaningless

* * * * *

PECULIAR SIGNIFICANCE OF SOME DREAMS

From practical experience it has been ascertained that if any one resumes his sleep after dreaming a dream, its good as well as bad effects are neutralized. Accordingly, after seeing a bad dream, one should renew one's sleep and thereby get rid of its influence.

The under-mentioned diseases are more or less incurable and death is certain to result from them, if the patient, in addition, dreams the following dreams during his or her illness :—

(1) Friendship with dogs in the case of *Chronic fever.*

(2) Friendship with monkeys in the case of *Phthisis.*

(3) Drinking water in the case of *Spermatorrhœa* and *Dyspepsia.*

(4) Drinking clarified butter or oil in the case of *Leprosy.*

(5) Growth of a tree on the head in the case of *Chronic Enlargement of Spleen* and the head diseases.

(6) Eating of cakes by the patient in the case of vomiting.

(7) Things mixed with turmeric in the case of *Asthma* and thirst diseases.

(8) Tasting of blood in the case of *Spontaneous Hæmorrhage* from the mouth.

One should not disbelieve the above statement. These particular dreams are invariably dreamt by the patients, when the diseases have advanced to an incurable stage and death-signs have become perceptible on their persons.

A SIMPLE TALK ABOUT MAGNETISM

Animal magnetism. Its meaning. How to attain it etc.

The animal magnetism had not been unknown to the Aryan saints (Rishis) before it was brought to light into the western world by Mesmer. They (Rishis) are of opinion that a very fine and subtle substance, finer and more subtle than ether, pervades this universe. The mutual attraction of the far-off planets and satellites is nothing but an atomic vibration of that finer substance which enables a living body to exercise such an enormous influence over another. This is known as animal magnetism.

Towards the latter part of the 19th century Albrecht Von Heller, a famous Physiologist demonstrated this view partially. He holds that there exists in human system a kind of subtle force, lying in a dormant state, which is made potent by the moving of our limbs. The famous scientist, A Von Humbolot, says that the action of this nerve force can be realised even from a distant.

Edward Von, the well-known scientist of the modern age, also supports this view. One might have ignored this theory had it been promulgated by Rishis (saints) or Ojhas (exorcists) but now can any one disregard this view as expressed by the western scholars endeavouring to study the high destiny of mankind with the help of the Aryan theory.

Thus it is seen that each individual possesses a hidden force. By the proper culture of which he can be a maker of his own destiny and can make his life beautiful and glorious or miserable. The process to attain this wonderful power is explained in the next chapter on Aryan Hypnotism. Now the success greatly depends upon the concentration of one's mind. One having no concentration of mind is doomed to failure as his internal senses are not fully developed by the proper exercise of breathing laws as given in this book.

This is the reason why some are successful and some in their efforts are unable to gain their ends after a protracted wilful confinement in the temple of a God or Goddess (a practice prevalent in India).

Magnetic sides of human body

Like the two poles of a magnet the right and the left sides of a man produce different results. The right side of every human being possesses the positive force while that of the left side the negative force. Thus the positive sides of two bodies, one side facing the other, as also the negative sides of two bodies, one side facing the other, have the repelling capacity like the two like poles of a magnet. The positive side of the body, when placed near or in contact with the negative side of another body, has the capacity of attraction as in the case of the unlike poles of a magnet. Thus the scientific theory has been deduced tha like poles repel and the unlike poles attract each other.

Acting upon the above magnetic principle in

human system the Aryans have laid down in the Hindu Scriptures the practice of placing the wife on the left side of her husband with a view to avoid or remove the bitterness of feeling or animosity, if there be any between the married couple. The Hindu sages can produce in mesmerism or exorcism different results in strict conformity with the principle of positive and negative forces in man. The Aryans used magnet for medical purposes in many complicated diseases. The reasonableness of which is not even denied by many English scientists of the age.

Natural Laws to win love, affection, favour and all kinds of grace etc.

By virtue of this power one organ of the body can perform the functions of another. Hardenhen says,—one of his disciples performed the function of his ears by the help of his navel ; some are seen reading by merely touching the books without the help of their eyes. Mr. Mole says that he saw a person reading by the help of the nip of his nose a book lying three feet apart with his eyes blindfolded or covered.

The practical seekers after this knowledge should read other books of eminent writers on this subject. Our intention is only to put the readers to the right directions. If they are really thirsty of it they will get it sooner or later. Now let us find out the ways and means to acquire this hidden force. Without entering into the detailed forms of diet essential for this purpose, let us only mention that the simple diet of vegetables, which is not highly spiced or rich, is greatly helpful in this matter; one is to dip a magnet in a glass of water and drink the water surcharged with the magnet. The control over mind and senses, concentration of mind and the acquirement of foresight are to be practised next.

Nature is vast and human knowledge is small, insignificant and limited. Why happiness brings on smile and misery tears and how out of a small drop of semen a beautiful and lovely baby is born? These are the Mysteries of Nature lying beyond the pale of human intelligence. Has any body found out the best reason for any strange phenomenon?

Why then should men cry down any thing as strange and impossible? Why should men give wide birth to, instead of perusing and thereby being charmed at the beauty of a subject, apparently difficult? The importance of knowledge is high.

THE ARYAN HYPNOTISM

At the request of many of my readers both far and near I am giving below some practical lessons on hypnotism which the Fakirs and sages of India daily perform in the broad streets. I am sure that a careful study of these methods will greatly benefit the readers in their voyage of life.

It should be well remembered that to get Jnana Kanda (true knowledge) one should totally forget that there are many substances, for, indeed, these things are but *Maya* (illusion). There is only one substance, and all this Universe exists in the *Paramatma* or the Universal Spirit; as in innumerable cups of water the sun may be reflected many times but the substance is the same, so there are innumerable persons in this world, like the cups of water, but the animating spirit like the Sun is one.

The *Paramatma* or the spirit exists in whatever was, is or will be, either formless or the formed.

That intelligence which is inherent in us comes from the Universal Spirit, and intelligence which we like to impart upon others are but one from the same luminous source i. e. the spirit.

All this Universe, seen or unseen, comes out of intelligence, and in putting the senses to sleep we but practise the concentrated power of the One.

The reader should always bear in mind that the Universe .comes out of *Maya* (illusion). The air came from the ether, the fire from the air, water from the fire, and the earth emanated from the water. This is the order of emanation. This is the verdict of the wise. Qualities exist in all things and many distinctions arise only in names and words, but to speak the truth there is no existence of any real difference. Therefore the things which we see around us will not exist but their manifestation, the Great and Glorious One, will exist up to the eternity.

In order to attain hypnotic power one should first of all learn to control his desires and will power, for, of a truth, the substance by which hypnotism is created is but *Maya* (illusion). Therefore he who has not completely shaken the bondage of both vice and virtue and fails to purify his body by the purification of *Nàdis*, will never attain the supernatural power to create illusion.

Now the easy methods are explained. One should attain success only through faith. He must undergo the practices with volcanic energy, adamantine labour and honest perseverance. If you are addicted to speak lies, to sensual pleasures, if you are a disbeliever, you will scarcely attain success in the line. He should at the beginning control over his animal propensities and thereby learn to concentrate his mind on one thing at a time forgetting everything for the time being. I earnestly request you to retire to a very solitary place

Easy process to attain supernatural power

where, surrounded by sweet scented odours, you may be in peace and solitude. Sit on the floor with legs crossed and begin to practise *Pránáyama.* Then with your heart full of praise and afire with faith, close the *Pingalá Nádi* (the right nostril) with your right thumb. You will then take in air through the *Idá Nádi* (the left nostril), and keep both the nostrils closed—thus suspending your breath—for as long a period as possible, and when you seem to be unable to hold the breath any longer, then breathe out through the right nostril slowly and not rapidly, stopping the left nostril with the left thumb.

When the breath has been completely exhaled, again draw the breath in the same manner through the *Pingalá,* and stop from breathing as long as you can and then give out the breath through the *Idá.* This practice is called the *Kumbhaka* (stopping of breath). It has been clearly explained elsewhere in this book. Please read the rules there carefully to master it. Perform at least 15 to 20 *Kumbhakas* daily four times,—at sunrise, at midday, at sunset and at midnight. This practice should be performed daily without any neglect.

This simple practice will purify the body. All defects in the system will be removed, sweet smell will come out of the body and the skin will become pure. It will impart good appetite with digestion greatly increased. He who performs this will attain sweet voice, indomitable courage and mighty energy and zeal for hard work.

By the continued practice you will be able to stem the tide of misfortunes and calamities in your voyage across the world's sea of sorrow and your power of magnetism will be greatly increased in such a way that you can make a man do what you desire. You will be free from all sorts of diseases, sorrow and pain.

It has been mentioned in the first stage that the practice of *Pránáyama* will increase the Will Power. The eastern adepts can hold the breath for four hours and over at a time. The novice should attempt to hold the breath for a little time at the beginning. Gradually he may increase the period.

After six months' practice of the *Pránáyáma* one should attempt this following exercise :—

How to cure disease by hypnotism

"Place the patient on his back. The patient should be nude or covered only with a very thin garment. He must lie on his back and place his arms at his side."

Now direct the patient to concentrate his whole attention within himself, even within his brain. Within the brain lies the mystic Mount *Kailash,* the seat of the thousand petaled lotus. Command the patient to affix his whole mind on the lotus within the brain. Tell him to imagine a hole in the centre of the head just between the eye-brows. Now bid him to inhale the breath and try to fill this space where dwells the lotus. He should give up the outward thoughts of any kind but should keep on inhaling his breath into the hole with one mind. Tell him that this thousand petaled lotus is a lump of radiant light and let him try to picture it as he fills the hole with his breath. This hole is called the pineal gland.

Now take your hands and gently pass them over the man's forehead down over the ears and side of his neck and slowly over the breast and just below the breast. This should be done first from the head and stop just below the breast and then pass your hands in the upward direction.

The hands must be passed very slowly and very gently over the breast each time and must not touch the body on any account. Before giving passes rub the palms of your two hands by friction against each other. As the passes are applied you should intensely exercise your will with one mind that the patient shall sleep. At the second movement you will pass your hands up to the navel, returning in the same way to the centre of the forehead. At the third movement you will bring the hands up to the neck, over the breast, past the navel and down the outside of the legs to the knees.

At the fourth time repeat the same process, and at the knees you will bring your hands to the inside of the leg down the inside of the calf, pass it over the instep outwards and out at the toes. The man will all this time be putting his breath and thought into the hole in the head. When the fourth movement will be on operation the patient will begin to close his eyes for sleep. At this stage you should place your hand on the forehead of the patient, even between the eye-brows, and speak to him with your mind.

If the patient is attacked with pains in his body, or if he suffers from cold, gout or rheumatism you should breathe deeply upon the part of the body which is unwell. Exhale your breath upon it thrice or four times with hot breath over the injured part. You should all along think within yourself with one mind " The patient is well."

If a man has watery fluxes in his body, if he suffers from fevers or congestions, you should breathe upon the painful part of the body from afar, so that the breath may be cold when it reaches the affected part of the patient. In this way many cures have been effected.

"It is, perhaps, not generally known that susceptible subjects may be readily hypnotised while engaged in an ordinary conversation and without being aware of the intention of the hypnotist. Such, however, I have found to be the case. For subjects who really desire to experience the sensation of being under the spell of the hypnotist, but who, from a feeling of timidity or bashfulness, refuse to allow themselves to be influenced, this method is preferable to all others ; and for the bigoted know-it-all, who wants to bet you ten dollars you can't hypnotise *him,*—you have all met him, it is the method *par excellence.*"

Indirect hypnotism

The mode of procedure is simplicity itself, and is as follows :—In the first place select an easy subject for your first experiment. Begin to converse with him at such a place and time that there will be nothing to disturb him or divert his mind. Then proceed the conversation on such a subject which you can talk in an interesting and attractive manner. At last secure his undivided attention, when the transition will be very easy from that state to an intense concentration of his mind. If you are an excellent talker, at which time a positive suggestion, given in a sudden and forceful manner, will soon bring him under your direct control.

"I have found it a capital plan to turn the conversation upon hypnotism, and after dwelling upon some of its beauties and wonders till, as was said above, his mind is concentrated, proceed, as if illustrating some foregoing statement, something like this:—"Now, for example,—when I say to you, you have forgotten your own name, you have forgotten it, and *you cannot say your name,* accompanying this suggestion with a straight, steady, authoritative look, right into his eyes, and a pass or two will probably strengthen the suggestion, and you have hypnotised him unawares."

Though the subject of hypnotism is dealt here not so clearly yet the readers will get a stepping stone to it. The regulation of breath will be carefully mastered first to have a thorough knowledge of this subject.

Smiling is the best aid to health

When you feel very tired, or very discouraged, or very depressed, or very indigent, go to your mirror and practise smiling. Smiling has not only a quite serious psychological influence, but actually the muscles brought into play as you smile stimulate parts of the brain and the whole Nervous System. We all covet youth and we may all achieve it. If we think depressing thoughts we do not look young. If we allow our mind to get beyond our control and worry, for example, we at once begin to show the passing years in our face, in our figure, and in every thing we do. Yet every human being is capable of controlling the body by the mind. Age is an infirmity. Your sorroundings are important. It is not necessary to be rich to have a beautiful home. It is stimulating when one is depressed to go out and buy something which will add to the beauty of the home or the individual.

Face massage is a very good thing, just as massage of the whole body is good. The pity is so few take it up, and that such ridiculous charges are made for it. Every woman cannot do it for herself.

No woman need have a bad complexion now. When a woman feels that her complexion is all right, she feels all right herself. It gives her confidence.

It is not true to say you can't help worrying. We can all help it. Say to yourself every morning, 'I fear nothing.' It is the same thing as saying, 'I will be young.' Visualize what you want, and if you do it long enough you will get.

Effects of Alcohol, root cause of many diseases

Alcohol has an affinity for protoplasm, the physical basis of life itself and this is the cause for the widely varied injuries that it always inflicts on the human system ; the brain-tissues, the nerves, the heart and blood-vessels, the stomach and the intestinal passages, the lympathic system, the kidneys and the liver may each and all become deranged and diseased by the habitual drinking of even small quantities of alcohol. The effects of drinking alcohol are not only confined to the drinkers but are carried on to his descendants of successive generations. All these may be compared to the eroding effect of a running river which has its greatest action on the part of its bank that is least resistant.

Thus alcohol circulating in the blood tends to attack whatever organ or tissues may be weakest and most susceptible. In its action on the brain, alcohol is an inhibitor and disturber of mental activity, not a promoter of it. Wine causes a loss of at least ten per cent in working efficiency. It is not only a poison, but aids the action of other poisons, such as those of contagion.

Alienists everywhere agree that alcohol is a most potent cause of insanity. Statistics show also that it is the chief cause of crimes, pauperism and creates a disastrous effect on the community. For every individual that dies prematurely of a disease directly due to alcohol, there are scores of individuals that suffer to a lesser degree from maladies which are wholly or partially of the same origin but which are not directly fatal. For every incapable who, weakened by alcohol, acknowledges defeat in the life-battle and openly seeks alms, there are scores of individuals that feel the pleasure of want in a greater or less degree because the money that might have supplied necessaries and luxuries has gone for drink. Yet they strive to hide their indigence.

A SIMPLE PROCESS OF THOUGHT—READING

Here is an ingenious figure-diversion which you might try on some of your friends. Suppose you are all seated round a fire, and then ask one of you to take a book, open it at random, and select a word. By means of the magic of figures, you will be able to tell the particular word selected.

The process you should follow is to ask your friend to select any word from the first ten lines and within ten from the end of the line.

Then ask him to double the number of the page, and multiply the product by 5 and add 20; then, to the sum thus obtained add the number of the line, and increase it by 5, and then multiply the sum by 10. To the total add the number of the word in the line, and from the sum subtract 250.

The last figure of the total will always be the number of the word, the next figure the number of the line, and the remaining figures the number of the page. To cite a concrete illustration : some one selects, say, the fifth word of the eighth line, on page 402.

Double the number of the page	804
Multiply by 5	4,020
Add 20	4,040
Add the number of the line (8)	4,048
Add 5	4,053
Multiply by 10	40,530
Add the number of the word in the line (5)	40,535
Subtract 250	40,285

He then tells you the answer is 40,285, which gives you (the last figure) the number of the word, 5 ; next (the next figure) the number of the line 8 ; and the remaining figures will be the number of the page, 402. Now take the book, open it at page 402, look at the fifth word in the eighth line, and—you have it !

* * * * *

* * * *

Look at the spacious firmament on high, and you will be struck with the mysteries of the Almighty Being who sees us all, but shuts Himself out from the sight of the world by wearing a fold of heaven and earth across His face. The mysteries of nature should stimulate us all the more to strive to catch a glimpse of Him at least with our inner eyes. Only materialism keeps us aloof from Him. We should enjoy this world, but let us not be wholly given to the transient pleasures of this material world. In whatever station of life we may be placed, let us ever have for our motto in life the noble counsels contained in the following verses :—

Be pure in heart and pure in mind,
And pure in movement every kind.
Keep it aloof from every thing,
Conquer, control and be a king.

Think yourself as heavenly soul,
Make it spacious up to the pole.
Teach your mind to chew the noble view,
B't none to touch it tho' pleasing and new.

Feel kingly grace and kill all weeds,
But live and move in too slow speeds.
Study Nature that moves around,
Grasp the idea, not the sound.

Have your mind attach'd to the heart,
Even in dangers not to part,
Parting is but fatal ruin—
Does benighted sun overshine ?

Heart and mind, and ether and sun,
Through time eternal always run.
Mark, mark the time—how it arises,
From where it comes, to where it goes.

What leaves behind and what it takes,
What symbol all along it makes.
Often beware of that it may not
Make you its victim while so thought.

THE END